Combinatorial Methods for String Processing

Combinatorial Methods for String Processing

Editor

Shunsuke Inenaga

MDPI • Basel • Beijing • Wuhan • Barcelona • Belgrade • Manchester • Tokyo • Cluj • Tianjin

Editor
Shunsuke Inenaga
Department of Informatics
Kyushu University
Fukuoka
Japan

Editorial Office
MDPI
St. Alban-Anlage 66
4052 Basel, Switzerland

This is a reprint of articles from the Special Issue published online in the open access journal *Algorithms* (ISSN 1999-4893) (available at: www.mdpi.com/journal/algorithms/special_issues/ String_Processing).

For citation purposes, cite each article independently as indicated on the article page online and as indicated below:

LastName, A.A.; LastName, B.B.; LastName, C.C. Article Title. *Journal Name* **Year**, *Volume Number*, Page Range.

ISBN 978-3-0365-2194-7 (Hbk)
ISBN 978-3-0365-2193-0 (PDF)

Contents

 **algorithms**

Article

Efficient Data Structures for Range Shortest Unique Substring Queries [†]

Paniz Abedin [1], Arnab Ganguly [2], Solon P. Pissis [3,4,*] and Sharma V. Thankachan [1]

[1]   Department of Computer Science, University of Central Florida, Orlando, FL 32816, USA;
      paniz@knights.ucf.edu (P.A.); sharma.thankachan@ucf.edu (S.V.T.)
[2]   Department of Computer Science, University of Wisconsin - Whitewater, Whitewater, WI 53190, USA;
      gangulya@uww.edu
[3]   Life Sciences and Health, CWI, 1098 XG Amsterdam, The Netherlands
[4]   Center for Integrative Bioinformatics, Vrije Universiteit, 1081 HV Amsterdam, The Netherlands
*   Correspondence: solon.pissis@cwi.nl
†   An early version of this paper appeared in the Proceedings of SPIRE 2019.

Received: 9 September 2020; Accepted: 29 October 2020; Published: 30 October 2020

Abstract: Let $T[1, n]$ be a string of length n and $T[i, j]$ be the substring of T starting at position i and ending at position j. A substring $T[i, j]$ of T is a repeat if it occurs more than once in T; otherwise, it is a unique substring of T. Repeats and unique substrings are of great interest in computational biology and information retrieval. Given string T as input, the Shortest Unique Substring problem is to find a shortest substring of T that does not occur elsewhere in T. In this paper, we introduce the range variant of this problem, which we call the Range Shortest Unique Substring problem. The task is to construct a data structure over T answering the following type of online queries efficiently. Given a range $[\alpha, \beta]$, return a shortest substring $T[i, j]$ of T with exactly one occurrence in $[\alpha, \beta]$. We present an $\mathcal{O}(n \log n)$-word data structure with $\mathcal{O}(\log_w n)$ query time, where $w = \Omega(\log n)$ is the word size. Our construction is based on a non-trivial reduction allowing for us to apply a recently introduced optimal geometric data structure [Chan et al., ICALP 2018]. Additionally, we present an $\mathcal{O}(n)$-word data structure with $\mathcal{O}(\sqrt{n} \log^{\epsilon} n)$ query time, where $\epsilon > 0$ is an arbitrarily small constant. The latter data structure relies heavily on another geometric data structure [Nekrich and Navarro, SWAT 2012].

Keywords: shortest unique substring; suffix tree; heavy-light decomposition; range queries; geometric data structures

---

1. Introduction

Finding regularities in strings is one of the main topics of combinatorial pattern matching and its applications [1]. Among the most well-studied types of string regularities is the notion of repeat. Let $T[1, n]$ be a string of length n. A substring $T[i, j]$ of T is called a repeat if it occurs more than once in T. The notion of unique substring is dual: it is a substring $T[i, j]$ of T that does not occur more than once in T. Computing repeats and unique substrings has applications in computational biology [2,3] and information retrieval [4,5].

In this paper, we are interested in the notion of shortest unique substring. All of the shortest unique substrings of string T can be computed in $\mathcal{O}(n)$ time using the suffix tree data structure [6,7]. Many different problems based on this notion have already been studied. Pei et al. [4] considered the following problem on the so-called position (or point) queries. Given a position i of T, return a shortest unique substring of T covering i. The authors gave an $\mathcal{O}(n^2)$-time and $\mathcal{O}(n)$-space algorithm, which finds the shortest unique substring covering every position of T. Since then, the problem has been revisited and optimal $\mathcal{O}(n)$-time algorithms have been presented by Ileri et al. [8] and Tsuruta et al. [9]. Several other variants of this problem have been investigated [10–19].

We introduce a natural generalization of the shortest unique substring problem. Specifically, our focus is on the range version of the problem, which we call the Range Shortest Unique Substring (rSUS) problem. The task is to construct a data structure over T to be able to answer the following type of online queries efficiently. Given a range $[\alpha, \beta]$, return a shortest substring $T[k, k + h - 1]$ of T with exactly one occurrence (starting position) in $[\alpha, \beta]$; i.e., $k \in [\alpha, \beta]$, there is no $k' \in [\alpha, \beta]$ ($k' \neq k$), such that $T[k, k + h - 1] = T[k', k' + h - 1]$, and h is minimal. Note that this substring, $T[k, k + h - 1]$, may end at a position $k + h - 1 > \beta$. Further note that there may be multiple shortest unique substrings.

Range queries are a classic data structure topic [20–22]. A range query $q = f(A, i, j)$ on an array of n elements over some set S, denoted by $A[1, n]$, takes two indices $1 \leq i \leq j \leq n$, a function f defined over arrays of elements of S, and outputs $f(A[i, j]) = f(A[i], \ldots, A[j])$. Range query data structures have also been specifically considered for strings [23–26]. For instance, in bioinformatics applications we are often interested in finding regularities in certain regions of a DNA sequence [27–31]. In the Range–LCP problem, defined by Amir et al. [23], the task is to construct a data structure over T to be able to answer the following type of online queries efficiently. Given a range $[\alpha, \beta]$, return $i, j \in [\alpha, \beta]$ such that the length of the longest common prefix of $T[i, n]$ and $T[j, n]$ is maximal among all pairs of suffixes within this range. The state of the art is an $\mathcal{O}(n)$-word data structure supporting $\mathcal{O}(\log^{\mathcal{O}(1)} n)$-time (polylogarithmic-time) queries [25] (see also [26,32]).

1.1. Main Problem and Main Results

An alphabet Σ is a finite nonempty set of elements called letters. We fix a string $T[1, n] = T[1] \cdots T[n]$ over Σ. The length of T is denoted by $|T| = n$. By $T[i, j] = T[i] \cdots T[j]$, we denote the substring of T starting at position i and ending at position j of T. We say that another string P has an occurrence in T or, more simply, that P occurs in T if $P = T[i, i + |P| - 1]$, for some i. Thus, we characterize an occurrence of P by its starting position i in T. A prefix of T is a substring of T of the form $T[1, i]$ and a suffix of T is a substring of T of the form $T[i, n]$.

We next formally define the main problem considered in this paper.

Problem rSUS
Preprocess: String $T[1, n]$.
Query: Range $[\alpha, \beta]$, where $1 \leq \alpha \leq \beta \leq n$.
Output: (p, ℓ) such that $T[p, p + \ell - 1]$ is a shortest string with exactly one occurrence in $[\alpha, \beta]$.

If $\alpha = \beta$ the answer $(\alpha, 1)$ is trivial. So, in the rest we assume that $\alpha < \beta$.

1 2 3 4 5 6 7 8 9 10 11 12 13 14 15 16 17 18 19 20 21

Example 1. *Given* $T = caabcaddaacaddaaaabac$ *and a query* $[\alpha, \beta] = [5, 16]$*, we need to find a shortest substring of* T *with exactly one occurrence in* $[5, 16]$*. The output here is* $(p, \ell) = (10, 2)$*, because* $T[10, 11] = ac$ *is the shortest substring of* T *with exactly one occurrence in* $[5, 16]$*.*

Our main results are summarized below. We consider the standard word-RAM model of computations with w-bit machine words, where $w = \Omega(\log n)$, for stating our results.

Theorem 1. *We can construct an* $\mathcal{O}(n \log n)$*-word data structure that can answer any rSUS query on* $T[1, n]$ *in* $\mathcal{O}(\log_w n)$ *time.*

Theorem 2. *We can construct an* $\mathcal{O}(n)$*-word data structure that can answer any rSUS query on* $T[1, n]$ *in* $\mathcal{O}(\sqrt{n} \log^\epsilon n)$ *time, where* $\epsilon > 0$ *is an arbitrarily small constant.*

1.2. Paper Organization

In Section 2, we prove Theorem 1 and, in Section 3, we prove Theorem 2. We conclude this paper in Section 4 with some future proposals. An early version of this paper appeared as [33]. When compared to that early version ([33]), Theorem 2 is new.

2. An $\mathcal{O}(n \log n)$-Word Data Structure

Our construction is based on ingredients, such as the suffix tree [7], heavy-light decomposition [34], and a geometric data structure for rectangle stabbing [35]. Let us start with some definitions.

Definition 1. *For a position $k \in [1, n]$ and $h \geq 1$, we define* $\mathsf{Prev}(k, h)$ *and* $\mathsf{Next}(k, h)$, *as follows:*

$$\mathsf{Prev}(k, h) = \max_{j}\{\{j < k \mid \mathsf{T}[k, k + h - 1] = \mathsf{T}[j, j + h - 1]\} \cup \{-\infty\}\}$$
$$\mathsf{Next}(k, h) = \min_{j}\{\{j > k \mid \mathsf{T}[k, k + h - 1] = \mathsf{T}[j, j + h - 1]\} \cup \{+\infty\}\}.$$

Intuitively, let x and y be the occurrences of $\mathsf{T}[k, k + h - 1]$ right before and right after the position k, respectively. Subsequently, $\mathsf{Prev}(k, h) = x$ and $\mathsf{Next}(k, h) = y$. If x (resp., y) does not exist, then $\mathsf{Prev}(k, h) = -\infty$ (resp., $\mathsf{Next}(k, h) = +\infty$).

Definition 2. *Let $k \in [a, b]$. We define $\lambda(a, b, k)$, as follows:*

$$\lambda(a, b, k) = \min\{h \mid \mathsf{Prev}(k, h) < a \textbf{ and } \mathsf{Next}(k, h) > b\}.$$

Intuitively, $\lambda(a, b, k)$ denotes the length of the shortest substring that starts at position k with exactly one occurrence in $[a, b]$.

Definition 3. *For a position $k \in [1, n]$, we define C_k, as follows:*

$$C_k = \{h > 1 \mid (\mathsf{Next}(k, h), \mathsf{Prev}(k, h)) \neq (\mathsf{Next}(k, h - 1), \mathsf{Prev}(k, h - 1))\} \cup \{1\}$$

Example 2. *(Running Example for Definition 3) Let* $\mathsf{T} = \overset{\text{1 2 3 4 5 6 7 8 9 10 11 12 13 14 15 16 17 18 19 20 21}}{caabcaddaacaddaaaabac}$ *and* $k =$ 10. *We have that* $(\mathsf{Next}(10, 1), \mathsf{Prev}(10, 1)) = (12, 9)$, $(\mathsf{Next}(10, 2), \mathsf{Prev}(10, 2)) = (20, -\infty)$, *and* $(\mathsf{Next}(10, 3), \mathsf{Prev}(10, 3)) = (+\infty, -\infty)$. *Thus,* $C_{10} = \{2, 3\} \cup \{1\} = \{1, 2, 3\}$.

Intuitively, C_k stores the set of candidate lengths for the shortest unique substrings starting at position k. We make the following observation.

Observation 1. $\lambda(a, b, k) \in C_k$, *for any* $1 \leq a \leq b \leq n$.

Example 3. *(Running Example for Observation 1) Let* $\mathsf{T} = \overset{\text{1 2 3 4 5 6 7 8 9 10 11 12 13 14 15 16 17 18 19 20 21}}{caabcaddaacaddaaaabac}$ *and* $k = 10$. *We have that* $C_{10} = \{1, 2, 3\}$. *For $a = 5$ and $b = 16$, $\lambda(5, 16, 10) = 2$, denoting substring* **ac**. *For $a = 5$ and $b = 20$, $\lambda(5, 20, 10) = 3$, denoting substring* **aca**.

The following combinatorial lemma is crucial for efficiency.

Lemma 1. $\sum_k |C_k| = \mathcal{O}(n \log n)$.

Proof. The proof of Lemma 1 is deferred to Section 2.1. $\square$

We are now ready to present our construction. By Observation 1, for a given query range $[\alpha, \beta]$, the answer (p, ℓ) that we are looking for is the pair (k, h) with the minimum h under the following conditions: $k \in [\alpha, \beta]$, $h \in C_k$, $\mathsf{Prev}(k, h) < \alpha$ and $\mathsf{Next}(k, h) > \beta$. Equivalently, (p, ℓ) is the pair (k, h) with the minimum h, such that $h \in C_k$, $\alpha \in (\mathsf{Prev}(k, h), k]$, and $\beta \in [k, \mathsf{Next}(k, h))$. We map each $h \in C_k$ into a weighted rectangle $R_{k,h}$ with weight h, which is defined as follows:

$$R_{k,h} = [\mathsf{Prev}(k, h) + 1, k] \times [k, \mathsf{Next}(k, h) - 1].$$

Let $\mathcal{R}$ be the set of all such rectangles, then the lowest weighted rectangle in $\mathcal{R}$ stabbed by the point (α, β) is $R_{p,\ell}$. In short, an rSUS query on $\mathsf{T}[1, n]$ with an input range $[\alpha, \beta]$ can be reduced to an equivalent top-1 rectangle stabbing query on a set $\mathcal{R}$ of rectangles with input point (α, β). In the 2-d Top-1 Rectangle Stabbing problem, we preprocess a set of weighted rectangles in 2-d, so that given a query point q the task is to report the largest (or lowest) weighted rectangles containing q [35]. Similarly, here, the task is to report the lowest weighted rectangle in $\mathcal{R}$ containing the point (α, β) (see Figure 1 for an illustration). By Lemma 1, we have that $|\mathcal{R}| = \mathcal{O}(n \log n)$. Therefore, by employing the optimal data structure for top-1 rectangle stabbing presented by Chan et al. [35], which takes $\mathcal{O}(|\mathcal{R}|)$-word space supporting $\mathcal{O}(\log_w |\mathcal{R}|)$-time queries, we arrive at the space-time trade-off of Theorem 1. This completes our construction.

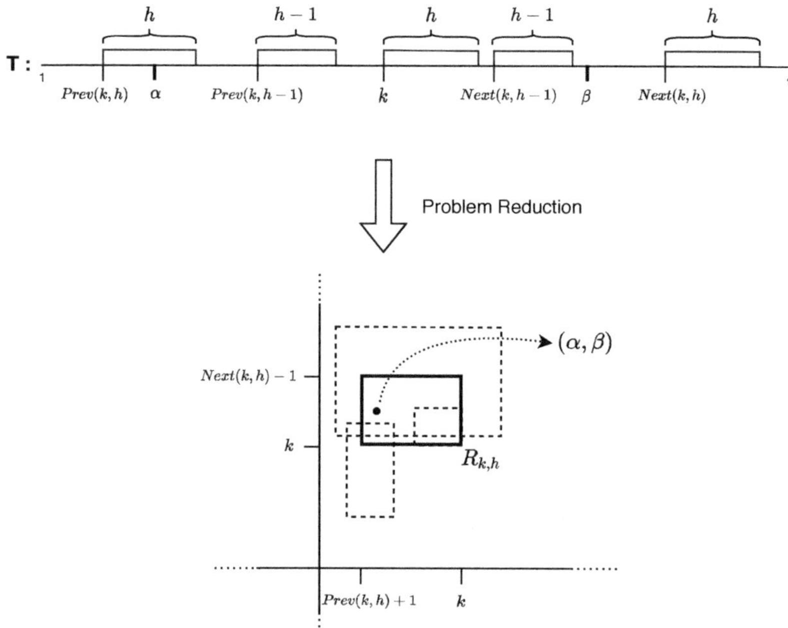

Figure 1. Illustration of the problem reduction: (k, h) is the output of the rSUS problem with query range $[\alpha, \beta]$, where $h = \lambda(\alpha, \beta, k) \in C_k$. $R_{k,h}$ is the lowest weighted rectangle in $\mathcal{R}$ containing the point (α, β).

2.1. Proof of Lemma 1

Let $\mathsf{lcp}(i, j)$ denote the length of the longest common prefix of the suffixes of T starting at positions i and j in T. Additionally, let S denote the set of all (x, y) pairs, such that $1 \leq x < y \leq n$ and $\mathsf{lcp}(x, y) > \mathsf{lcp}(x, z)$, for all $z \in [x+1, y-1]$. The proof can be broken down into Lemma 2 and Lemma 3.

Lemma 2. $\sum_k |C_k| = \mathcal{O}(|S|)$.

Proof. Let us fix a position k. Let

$$C_k' = \{h > 1 \mid \mathsf{Prev}(k,h) \neq \mathsf{Prev}(k,h-1)\}$$

$$C_k'' = \{h > 1 \mid \mathsf{Next}(k,h) \neq \mathsf{Next}(k,h-1)\}.$$

Clearly we have that $C_k = C_k' \cup C_k'' \cup \{1\}$.

The following statements can be deduced by a simple contradiction argument:

1. Let $i = \text{Prev}(k, h) \neq -\infty$, where $h \in C'_k$, then $i = \text{Prev}(k, \text{lcp}(i, k))$
2. Let $j = \text{Next}(k, h) \neq \infty$, where $h \in C''_k$, then $j = \text{Next}(k, \text{lcp}(k, j))$.

Figure 2 illustrates the proof for the first statement. The second one can be proved in a similar fashion.

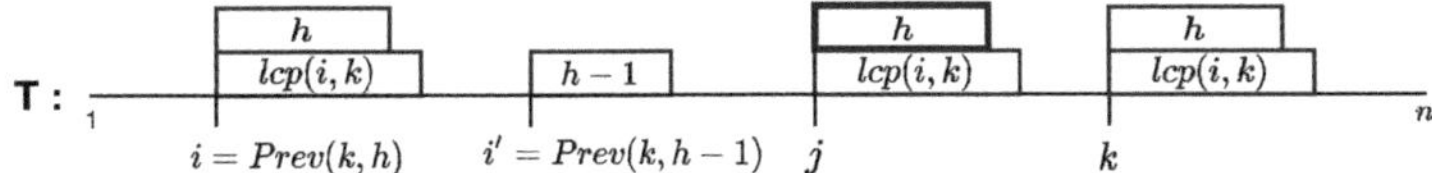

Figure 2. Let $h \in C'_k$ and $i = \text{Prev}(k, h)$. By contradiction, assume that there exists $j \in (i, k)$ such that $j = \text{Prev}(k, \text{lcp}(i, k))$. Since $h \leq \text{lcp}(i, k)$, $\mathsf{T}[j, j + h - 1] = \mathsf{T}[k, k + h - 1]$. This is a contradiction with $i = \text{Prev}(k, h)$. Thus, $i = \text{Prev}(k, \text{lcp}(i, k))$.

Clearly, $|C'_k|$ is proportional to the number of (i, k) pairs, such that $\text{lcp}(i, k) \neq 0$ and $i = \text{Prev}(k, \text{lcp}(i, k))$. Similarly, $|C''_k|$ is proportional to the number of (k, j) pairs, such that $\text{lcp}(k, j) \neq 0$ and $j = \text{Next}(k, \text{lcp}(k, j))$. Therefore, $\sum_k |C_k|$ is proportional to the number of (x, y) pairs, such that $\text{lcp}(x, y) \neq 0$ and $\text{lcp}(x, y) > \text{lcp}(x, z)$, for all $z \in [x + 1, y - 1]$. This completes the proof of Lemma 2. $\square$

Lemma 3. $|S| = \mathcal{O}(n \log n)$.

Proof. Consider the suffix tree data structure of string $\mathsf{T}[1, n]$, which is a compact trie of the n suffixes of T appended with a letter $\$ \notin \Sigma$ [7]. This suffix tree consists of $n + 1$ leaves (one for each suffix of T) and at most n internal nodes. The edges are labeled with substrings of T. Let u be the lowest common ancestor of the leaves corresponding to the strings $\mathsf{T}[x, n]\$$ and $\mathsf{T}[y, n]\$$. Subsequently, the concatenation of the edge labels on the path from the root to u is exactly the longest common prefix of $\mathsf{T}[x, n]\$$ and $\mathsf{T}[y, n]\$$. For any node u, we denote, by $\text{size}(u)$, the total number of leaf nodes of the subtree rooted at u.

We decompose the nodes in the suffix tree into light and heavy nodes. The root node is light and for any internal node, exactly one child is heavy. Specifically, the heavy child is the one having the largest number of leaves in its subtree (ties are broken arbitrarily). All other children are light. This tree decomposition is known as heavy-light decomposition. We have the following critical observation. Any path from the root to a leaf node contains many nodes; however, the number of light nodes is at most $\log n$ [34,36]. Additionally, corresponding to the $n + 1$ leaves of the suffix tree, there are $n + 1$ paths from the root to the leaves. Therefore, the sum of subtree sizes over all light nodes is $\mathcal{O}(n \log n)$.

We are now ready to complete the proof. Let $S_u \subseteq S$ denote the set of pairs (x, y), such that the lowest common ancestor of the leaves corresponding to suffixes $\mathsf{T}[x, n]\$$ and $\mathsf{T}[y, n]\$$ is u. Clearly, the paths from the root to the leaves that correspond to suffixes $\mathsf{T}[x, n]\$$ and $\mathsf{T}[y, n]\$$ pass from two distinct children of node u and then at least one of the two must be a light node. There are two cases. In the first case, both leaves are under the light children. In the second case, one leaf is under a light child and the other is under the heavy child. In both cases, we have at least one leaf under a light node. If we fix the leaf that is under the light node, we can enumerate the total number of pairs based on the subtree size of the light nodes. Therefore, $|S_u|$ is at most twice the sum of $\text{size}(\cdot)$ over all light children of u. Since $|S| = \sum_u |S_u|$, we can bound $|S|$ by the sum of $\text{size}(\cdot)$ over all light nodes in the suffix tree, which is $\mathcal{O}(n \log n)$. This completes the proof of Lemma 3. $\square$

3. An $\mathcal{O}(n)$-Word Data Structure

This section is dedicated to proving Theorem 2. For simplicity, we only focus on the computation of the length ℓ of the output (p, ℓ).

Let SA be the suffix array of string T of length n, which is a permutation of $\{1, \cdots, n\}$, such that $\text{SA}[i] = j$ if $\mathsf{T}[j, n]$ is the ith lexicographically smallest suffix of T [37]. Further, let SA^{-1} be the *inverse suffix array* of string T of length n, which is a permutation of $\{1, \cdots, n\}$, such that $SA^{-1}[SA[i]] = i$. Moreover, SA of T can be constructed in linear time and space [38,39].

We observe that an $\mathcal{O}(\beta - \alpha + 1)$-time solution is straightforward with the aid of the suffix tree of T as follows. First, identify those leaves corresponding to the suffixes starting within $[\alpha, \beta]$ using the inverse suffix array of T and mark them. Subsequently, for each marked leaf, identify its lowest ancestor node (and double mark it), such that a marked neighbor is also under it. This can be done via at most two $\mathcal{O}(1)$-time Lowest Common Ancestor (LCA) queries over the suffix tree of T while using $\mathcal{O}(n)$ additional space [22]. Afterwards, find the minimum over the string-depth of all double-marked nodes, add 1 to it, and report it as the length ℓ. The correctness is readily verified.

We employ the above procedure when $\beta - \alpha + 1 < 3\Delta$, where Δ is a parameter to be set later. We now consider the case when $\beta - \alpha + 1 \geq 3\Delta$. Note that ℓ is the smallest element in $S^* = \{\lambda(\alpha, \beta, k) \mid k \in [\alpha, \beta]\}$. Let α' be the smallest number after α and β' be the largest number before β, such that α' and β' are multiples of Δ. Subsequently, S^* can be written as the union of $S' = \{\lambda(\alpha, \beta, k) \mid k \in [\alpha, \alpha' - 1] \cup [\beta' + 1, \beta]\}$ and $S'' = \{\lambda(\alpha, \beta, k) \mid k \in [\alpha', \beta']\}$. Furthermore, S'' can be written as $S_1'' \cup S_2'' \cup S_3''$, where

- $S_1'' = \{h = \lambda(\alpha, \beta, k) \mid k \in [\alpha', \beta'], \mathsf{Prev}(k, h) \in [\alpha' - \Delta, \alpha - 1]\}$
- $S_2'' = \{h = \lambda(\alpha, \beta, k) \mid k \in [\alpha', \beta'], \mathsf{Next}(k, h) \in [\beta + 1, \beta' + \Delta]\}$
- $S_3'' = \{h = \lambda(\alpha, \beta, k) \mid k \in [\alpha', \beta'], \mathsf{Prev}(k, h) \in (-\infty, \alpha' - \Delta - 1], \mathsf{Next}(k, h) \in [\beta' + \Delta + 1, \infty)\}$.

Our construction is based on a solution to the Orthogonal Range Predecessor/Successor in 2-d problem. A set $\mathcal{P}$ of n points in an $[1, n] \times [1, n]$ grid can be preprocessed into a linear-space data structure, such that the following queries can be answered in $\mathcal{O}(\log^\epsilon n)$ time per query [40]:

- $\mathsf{ORQ}([x', x''], [-\infty, y'']) = \arg\max_j\{(i, j) \in \mathcal{P} \cap [x', x''] \times [-\infty, y'']\}$
- $\mathsf{ORQ}([-\infty, x''], [y', y'']) = \arg\max_i\{(i, j) \in \mathcal{P} \cap [-\infty, x''] \times [y', y'']\}$
- $\mathsf{ORQ}([x', x''], [y', +\infty]) = \arg\min_j\{(i, j) \in \mathcal{P} \cap [x', x''] \times [y', +\infty]\}$
- $\mathsf{ORQ}([x', +\infty], [y', y'']) = \arg\min_i\{(i, j) \in \mathcal{P} \cap [x', +\infty] \times [y', y'']\}$.

We next show how to maintain additional structures, so that the smallest element in each of the above sets can be efficiently computed and, thus, the smallest among them can be reported as ℓ.

- Computing the Smallest Element in S': for each $k \in [\alpha, \alpha' - 1] \cup [\beta' + 1, \beta]$, we compute $\lambda(\alpha, \beta, k)$ and report the smallest among them. We handle each $\lambda(\alpha, \beta, k)$ query in time $\mathcal{O}(\log^\epsilon n)$, as follows: first find the leaf corresponding to the string position k in the suffix tree of T, then the last (resp., first) leaf on its left (resp., right) side, such that the string position x (resp., y) corresponding to it is in $[\alpha, \beta]$, and report $1 + \max\{\mathsf{lcp}(k, x), \mathsf{lcp}(k, y)\}$. To efficiently enable the computation of x (resp., y), we preprocess the suffix array into an $\mathcal{O}(n)$-word data structure that can answer orthogonal range predecessor (resp., successor) queries in $\mathcal{O}(\log^\epsilon n)$ time [40].
- Computing the Smallest Element in S_1'': for each $r \in [\alpha' - \Delta, \alpha - 1]$, we compute the smallest element in $\{h = \lambda(\alpha, \beta, k) \mid k \in [\alpha', \beta'], \mathsf{Prev}(k, h) = r\}$ and report the smallest among them. The procedure is the following: find the leaf corresponding to the string position r in the suffix tree of T and the last (resp., first) leaf on its left (resp., right) side, such that its corresponding string position x (resp., y) is in $[\alpha', \beta']$ (via orthogonal range successor/predecessor queries as earlier). Subsequently, $t = \max\{\mathsf{lcp}(r, x), \mathsf{lcp}(r, y)\}$ is the length of the longest prefix of $T[r, n]$ with an occurrence d in $[\alpha', \beta']$. However, we need to verify whether occurrence d is unique and its $\mathsf{Prev}(d, t) = r$. For this, find the two leftmost occurrences of $T[r, r + t - 1]$ after r, denoted by x' and y' ($x' < y'$), via two orthogonal range successor queries. If y' does not exist, set $y' = +\infty$. Then report $\lambda(\alpha, \beta, d)$ if $\alpha' \leq x' \leq \beta' < y'$. Otherwise, report $+\infty$.
- Computing the Smallest Element in S_2'': for each $r \in [\beta + 1, \beta' + \Delta]$, we compute the smallest element in $\{h = \lambda(\alpha, \beta, k) \mid k \in [\alpha', \beta'], \mathsf{Next}(k, h) = r\}$ and report the smallest among them. The procedure is analogous to that of S_1''; i.e., find the length t of the longest prefix of $T[r, n]$ with an occurrence d in $[\alpha', \beta']$. Then, find the two rightmost occurrences of $T[r, r + t - 1]$ before r,

denoted by x' and y' ($x' < y'$), via two orthogonal range successor queries. If x' does not exist, set $x' = +\infty$. Subsequently, report $\lambda(\alpha, \beta, d)$ if $x' < \alpha' \leq y' \leq \beta'$. Otherwise, report $+\infty$.

- Computing the Smallest Element in S''_3: the set S''_3 can be written as $\{\lambda(\alpha' - \Delta, \beta' + \Delta, k) \mid k \in [\alpha', \beta']\}$, which is now dependent only on α', β' and Δ. Therefore, our idea is to pre-compute and explicitly store the minimum element in $\{\lambda(a - \Delta, b + \Delta, k) \mid k \in [a, b]\}$ for all (a, b) pairs, where both a and b are multiples of Δ, and for that the desired answer can be retrieved in constant time. The additional space needed is $\mathcal{O}((n/\Delta)^2)$.

We set $\Delta = \lfloor \sqrt{n} \rfloor$. The total space is then $\mathcal{O}(n)$ and total time is $\mathcal{O}(\Delta \log^\epsilon n) = \mathcal{O}(\sqrt{n} \log^\epsilon n)$. Therefore, we arrive at Theorem 2.

4. Final Remarks

We introduced the Range Shortest Unique Substring (rSUS) problem, the range variant of the Shortest Unique Substring problem. We presented a $\mathcal{O}(n \log n)$-word data structure with $\mathcal{O}(\log_w n)$ query time, where $w = \Omega(\log n)$ is the word size, for this problem. We also presented a $\mathcal{O}(n)$-word data structure with $\mathcal{O}(\sqrt{n} \log^\epsilon n)$ query time, where $\epsilon > 0$ is an arbitrarily small constant.

We leave the following related questions unanswered:

1. Can we design an $\mathcal{O}(n)$-word data structure for the rSUS problem with polylogarithmic query time?
2. Can we design an efficient solution for the k mismatches/edits variation of the rSUS problem, perhaps using the framework of [41]?
3. Can our reduction from Section 2 be extended to other types of string regularities, such as shortest absent words [42]?

Author Contributions: The initial draft of this paper was written by P.A. and S.V.T. S.P.P. wrote the introduction section and modified the whole draft. A.G. reviewed and edited the paper. All authors contributed to the manuscript equally by providing critical feedback and helping with the presentation. All authors have read and agreed to the published version of the manuscript.

Funding: This research was supported in part by the U.S. National Science Foundation (NSF) under CCF-1703489.

Conflicts of Interest: The authors declare no conflict of interest.

References

1. Lothaire, M. *Applied Combinatorics on Words*; Cambridge University Press: Cambridge, UK, 2005.
2. Schleiermacher, C.; Ohlebusch, E.; Stoye, J.; Choudhuri, J.V.; Giegerich, R.; Kurtz, S. REPuter: The manifold applications of repeat analysis on a genomic scale. *Nucleic Acids Res.* **2001**, *29*, 4633–4642, doi:10.1093/nar/29.22.4633.
3. Haubold, B.; Pierstorff, N.; Möller, F.; Wiehe, T. Genome comparison without alignment using shortest unique substrings. *BMC Bioinform.* **2005**, *6*, 123, doi:10.1186/1471-2105-6-123.
4. Pei, J.; Wu, W.C.; Yeh, M. On shortest unique substring queries. In Proceedings of the 29th IEEE International Conference on Data Engineering (ICDE 2013), Brisbane, Australia, 8–12 April 2013; pp. 937–948, doi:10.1109/ICDE.2013.6544887.
5. Khmelev, D.V.; Teahan, W.J. A Repetition Based Measure for Verification of Text Collections and for Text Categorization. In Proceedings of the 26th Annual International ACM SIGIR Conference on Research and Development in Informaion Retrieval, Toronto, Canada, 28 July–1 August 2003; pp. 104–110, doi:10.1145/860435.860456.
6. Gusfield, D. *Algorithms on Strings, Trees, and Sequences: Computer Science and Computational Biology*; Cambridge University Press: Cambridge, UK, 1997, doi:10.1017/cbo9780511574931.
7. Weiner, P. Linear Pattern Matching Algorithms. In Proceedings of the 14th Annual Symposium on Switching and Automata Theory (SWAT 1973), Iowa City, IA, USA, 15–17 October 1973; pp. 1–11, doi:10.1109/SWAT.1973.13.

8. Ileri, A.M.; Külekci, M.O.; Xu, B. Shortest Unique Substring Query Revisited. In Proceedings of the Combinatorial Pattern Matching—25th Annual Symposium (CPM 2014), Moscow, Russia, 16–18 June 2014; pp. 172–181, doi:10.1007/978-3-319-07566-2_18.

9. Tsuruta, K.; Inenaga, S.; Bannai, H.; Takeda, M. Shortest Unique Substrings Queries in Optimal Time. In Proceedings of the 40th International Conference on Current Trends in Theory and Practice of Computer Science, Nový Smokovec, Slovakia, 26–29 January 2014; pp. 503–513, doi:10.1007/978-3-319-04298-5_44.

10. Abedin, P.; Külekci, M.O.; V Thankachan, S. A Survey on Shortest Unique Substring Queries. *Algorithms* **2020**, *13*, 224.

11. Allen, D.R.; Thankachan, S.V.; Xu, B. A Practical and Efficient Algorithm for the k-mismatch Shortest Unique Substring Finding Problem. In Proceedings of the 2018 ACM International Conference on Bioinformatics, Computational Biology, and Health Informatics (BCB 2018), Washington, DC, USA, 29 August–1 September 2018; pp. 428–437, doi:10.1145/3233547.3233564.

12. Ganguly, A.; Hon, W.; Shah, R.; Thankachan, S.V. Space-Time Trade-Offs for the Shortest Unique Substring Problem. In Proceedings of the 27th International Symposium on Algorithms and Computation (ISAAC), Sydney, Australia, 12–14 December 2016; pp. 34:1–34:13, doi:10.4230/LIPIcs.ISAAC.2016.34.

13. Ganguly, A.; Hon, W.; Shah, R.; Thankachan, S.V. Space-time trade-offs for finding shortest unique substrings and maximal unique matches. *Theor. Comput. Sci.* **2017**, *700*, 75–88, doi:10.1016/j.tcs.2017.08.002.

14. Inoue, H.; Nakashima, Y.; Mieno, T.; Inenaga, S.; Bannai, H.; Takeda, M. Algorithms and combinatorial properties on shortest unique palindromic substrings. *J. Discret. Algorithms* **2018**, *52*, 122–132, doi:10.1016/j.jda.2018.11.009.

15. Hon, W.; Thankachan, S.V.; Xu, B. In-place algorithms for exact and approximate shortest unique substring problems. *Theor. Comput. Sci.* **2017**, *690*, 12–25, doi:10.1016/j.tcs.2017.05.032.

16. Mieno, T.; Inenaga, S.; Bannai, H.; Takeda, M. Shortest Unique Substring Queries on Run-Length Encoded Strings. In Proceedings of the 41st International Symposium on Mathematical Foundations of Computer Science MFCS, Kraków, Poland, 22–26 August 2016; pp. 69:1–69:11, doi:10.4230/LIPIcs.MFCS.2016.69.

17. Schultz, D.W.; Xu, B. On k-Mismatch Shortest Unique Substring Queries Using GPU. In Proceedings of the 14th International Symposium, Bioinformatics Research and Applications, Beijing, China, 8–11 June 2018, pp. 193–204, doi:10.1007/978-3-319-94968-0_18.

18. Mieno, T.; Köppl, D.; Nakashima, Y.; Inenaga, S.; Bannai, H.; Takeda, M. Compact Data Structures for Shortest Unique Substring Queries. In Proceedings of the 26th International Symposium, String Processing and Information Retrieval, Segovia, Spain, 7–9 October 2019; pp. 107–123, doi:10.1007/978-3-030-32686-9_8.

19. Watanabe, K.; Nakashima, Y.; Inenaga, S.; Bannai, H.; Takeda, M. Shortest Unique Palindromic Substring Queries on Run-Length Encoded Strings. In Proceedings of the 30th International Workshop Combinatorial Algorithms, Pisa, Italy, 23–25 July 2019; pp. 430–441, doi:10.1007/978-3-030-25005-8_35.

20. Yao, A.C. Space-time Tradeoff for Answering Range Queries (Extended Abstract). In Proceedings of the Fourteenth Annual ACM Symposium on Theory of Computing (STOC '82), San Francisco, CA, USA, 5–7 May 1982; pp. 128–136, doi:10.1145/800070.802185.

21. Berkman, O.; Vishkin, U. Recursive Star-Tree Parallel Data Structure. *SIAM J. Comput.* **1993**, *22*, 221–242, doi:10.1137/0222017.

22. Bender, M.A.; Farach-Colton, M. The LCA Problem Revisited. In Proceedings of the 4th Latin American Symposium, LATIN 2000: Theoretical Informatics, Punta del Este, Uruguay, 10–14 April 2000; pp. 88–94, doi:10.1007/10719839_9.

23. Amir, A.; Apostolico, A.; Landau, G.M.; Levy, A.; Lewenstein, M.; Porat, E. Range LCP. *J. Comput. Syst. Sci.* **2014**, *80*, 1245–1253, doi:10.1016/j.jcss.2014.02.010.

24. Amir, A.; Lewenstein, M.; Thankachan, S.V. Range LCP Queries Revisited. In Proceedings of the 22nd International Symposium, String Processing and Information Retrieval, London, UK, 1–4 September 2015; pp. 350–361, doi:10.1007/978-3-319-23826-5_33.

25. Abedin, P.; Ganguly, A.; Hon, W.; Nekrich, Y.; Sadakane, K.; Shah, R.; Thankachan, S.V. A Linear-Space Data Structure for Range-LCP Queries in Poly-Logarithmic Time. In Proceedings of the 24th International Conference, Computing and Combinatorics, Qing Dao, China, 2–4 July 2018; pp. 615–625, doi:10.1007/978-3-319-94776-1_51.

26. Ganguly, A.; Patil, M.; Shah, R.; Thankachan, S.V. A Linear Space Data Structure for Range LCP Queries. *Fundam. Inform.* **2018**, *163*, 245–251, doi:10.3233/FI-2018-1741.

27. Pissis, S.P. MoTeX-II: structured MoTif eXtraction from large-scale datasets. *BMC Bioinform.* **2014**, *15*, 235, doi:10.1186/1471-2105-15-235.
28. Almirantis, Y.; Charalampopoulos, P.; Gao, J.; Iliopoulos, C.S.; Mohamed, M.; Pissis, S.P.; Polychronopoulos, D. On avoided words, absent words, and their application to biological sequence analysis. *Algorithms Mol. Biol.* **2017**, *12*, 5:1–5:12, doi:10.1186/s13015-017-0094-z.
29. Ayad, L.A.K.; Pissis, S.P.; Polychronopoulos, D. CNEFinder: finding conserved non-coding elements in genomes. *Bioinformatics* **2018**, *34*, i743–i747, doi:10.1093/bioinformatics/bty601.
30. Iliopoulos, C.S.; Mohamed, M.; Pissis, S.P.; Vayani, F. Maximal Motif Discovery in a Sliding Window. In Proceedings of the 25th International Symposium, String Processing and Information Retrieval, Lima, Peru, 9–11 October 2018; pp. 191–205, doi:10.1007/978-3-030-00479-8_16.
31. Almirantis, Y.; Charalampopoulos, P.; Gao, J.; Iliopoulos, C.S.; Mohamed, M.; Pissis, S.P.; Polychronopoulos, D. On overabundant words and their application to biological sequence analysis. *Theor. Comput. Sci.* **2019**, *792*, 85–95, doi:10.1016/j.tcs.2018.09.011.
32. Matsuda, K.; Sadakane, K.; Starikovskaya, T.; Tateshita, M. Compressed Orthogonal Search on Suffix Arrays with Applications to Range LCP. In Proceedings of the 31st Annual Symposium on Combinatorial Pattern Matching, Copenhagen, Denmark, 17–19 June 2020; pp. 23:1–23:13, doi:10.4230/LIPIcs.CPM.2020.23.
33. Abedin, P.; Ganguly, A.; Pissis, S.P.; Thankachan, S.V. Range Shortest Unique Substring Queries. In Proceedings of the 26th International Symposium, String Processing and Information Retrieval, Segovia, Spain, 7–9 October 2019; pp. 258–266, doi:10.1007/978-3-030-32686-9_18.
34. Sleator, D.D.; Tarjan, R.E. A Data Structure for Dynamic Trees. In Proceedings of the 13th Annual ACM Symposium on Theory of Computing, Milwaukee, WI, USA, 11–13 May 1981; pp. 114–122, doi:10.1145/800076.802464.
35. Chan, T.M.; Nekrich, Y.; Rahul, S.; Tsakalidis, K. Orthogonal Point Location and Rectangle Stabbing Queries in 3-d. In Proceedings of the 45th International Colloquium on Automata, Languages, and Programming, Prague, Czech Republic, 9–13 July 2018; pp. 31:1–31:14, doi:10.4230/LIPIcs.ICALP.2018.31.
36. Harel, D.; Tarjan, R.E. Fast Algorithms for Finding Nearest Common Ancestors. *SIAM J. Comput.* **1984**, *13*, 338–355, doi:10.1137/0213024.
37. Manber, U.; Myers, E.W. Suffix Arrays: A New Method for On-Line String Searches. *SIAM J. Comput.* **1993**, *22*, 935–948, doi:10.1137/0222058.
38. Farach, M. Optimal Suffix Tree Construction with Large Alphabets. In Proceedings of the 38th Annual Symposium on Foundations of Computer Science (FOCS '97), Miami Beach, FL, USA, 19–22 October 1997; pp. 137–143, doi:10.1109/SFCS.1997.646102.
39. Kärkkäinen, J.; Sanders, P.; Burkhardt, S. Linear work suffix array construction. *J. ACM* **2006**, *53*, 918–936, doi:10.1145/1217856.1217858.
40. Nekrich, Y.; Navarro, G. Sorted Range Reporting. In Proceedings of the 13th Scandinavian Symposium and Workshops (SWAT 2012), Helsinki, Finland, 4–6 July 2012; pp. 271–282, doi:10.1007/978-3-642-31155-0_24.
41. Thankachan, S.V.; Aluru, C.; Chockalingam, S.P.; Aluru, S. Algorithmic Framework for Approximate Matching Under Bounded Edits with Applications to Sequence Analysis. In Proceedings of the 22nd Annual International Conference, Research in Computational Molecular Biology (RECOMB 2018), Paris, France, 21–24 April 2018; pp. 211–224, doi:10.1007/978-3-319-89929-9_14.
42. Barton, C.; Héliou, A.; Mouchard, L.; Pissis, S.P. Linear-time computation of minimal absent words using suffix array. *BMC Bioinform.* **2014**, *15*, 388, doi:10.1186/s12859-014-0388-9.

Publisher's Note: MDPI stays neutral with regard to jurisdictional claims in published maps and institutional affiliations.

 *algorithms*

Article

Computing Maximal Lyndon Substrings of a String

Frantisek Franek [1],[†] and Michael Liut [2],[*],[†] 

1   Department of Computing and Software, McMaster University, Hamilton, ON L8S 4K1, Canada;
    franek@mcmaster.ca
2   Department of Mathematical and Computational Sciences, University of Toronto Mississauga,
    Mississauga, ON L5L 1C6, Canada
*   Correspondence: michael.liut@utoronto.ca
†   These authors contributed equally to this work.

Received: 23 September 2020; Accepted: 10 November 2020; Published: 12 November 2020

Abstract: There are two reasons to have an efficient algorithm for identifying all right-maximal Lyndon substrings of a string: firstly, Bannai et al. introduced in 2015 a linear algorithm to compute all runs of a string that relies on knowing all right-maximal Lyndon substrings of the input string, and secondly, Franek et al. showed in 2017 a linear equivalence of sorting suffixes and sorting right-maximal Lyndon substrings of a string, inspired by a novel suffix sorting algorithm of Baier. In 2016, Franek et al. presented a brief overview of algorithms for computing the Lyndon array that encodes the knowledge of right-maximal Lyndon substrings of the input string. Among those presented were two well-known algorithms for computing the Lyndon array: a quadratic in-place algorithm based on the iterated Duval algorithm for Lyndon factorization and a linear algorithmic scheme based on linear suffix sorting, computing the inverse suffix array, and applying to it the *next smaller value* algorithm. Duval's algorithm works for strings over any ordered alphabet, while for linear suffix sorting, a constant or an integer alphabet is required. The authors at that time were not aware of Baier's algorithm. In 2017, our research group proposed a novel algorithm for the Lyndon array. Though the proposed algorithm is linear in the average case and has $O(n \log(n))$ worst-case complexity, it is interesting as it emulates the fast Fourier algorithm's recursive approach and introduces τ-reduction, which might be of independent interest. In 2018, we presented a linear algorithm to compute the Lyndon array of a string inspired by Phase I of Baier's algorithm for suffix sorting. This paper presents the theoretical analysis of these two algorithms and provides empirical comparisons of both of their C++ implementations with respect to the iterated Duval algorithm.

Keywords: combinatorics on words; string algorithms; regularities in strings; suffix sorting; Lyndon substrings; Lyndon arrays; right-maximal Lyndon substrings; tau-reduction algorithm; Baier's sort algorithm; iterative Duval algorithm

---

1. Introduction

In *combinatorics on words*, Lyndon words play a very important role. *Lyndon words*, a special case of Hall words, were named after Roger Lyndon, who was looking for a suitable description of the generators of free Lie algebras [1]. Despite their humble beginnings, to date, Lyndon words have facilitated many applications in mathematics and computer science, some of which are: constructing de Bruijin sequences, constructing bases in free Lie algebras, finding the lexicographically smallest or largest substring in a string, and succinct prefix matching of highly periodic strings; see Marcus et al. [2], and the informative paper by Berstel and Perrin [3] and the references therein.

The pioneering work on Lyndon decomposition was already introduced by Chen, Fox, and Lyndon in [4]. The *Lyndon decomposition theorem* is not explicitly stated there; nevertheless, it follows from the work presented there.

Theorem 1 (Lyndon decomposition theorem, Chen+Fox+Lyndon, 1958). *For any word* x, *there are unique Lyndon words* $u_1, ..., u_k$ *so that* $u_{i+1} \prec u_i$ *for any* $1 \leq i < k$, *and* $x = u_1 u_2 \ldots u_k$, *where* $\prec$ *denotes the lexicographic ordering.*

As there exists a bijection between Lyndon words over an alphabet of cardinality k and irreducible polynomials over F_k [5], many results are known about this factorization: the average number of factors, the average length of the longest factor [6] and of the shortest [7]. Several algorithms deal with Lyndon factorization. Duval gave in [8] an elegant algorithm that computes, in linear time and in-place, the factorization of a word into Lyndon words. More about its implementation can be found in [9]. In [10], Fredricksen and Maiorana presented an algorithm generating all Lyndon words up to a given length in lexicographical order. This algorithm runs in a constant average time.

Two of the latest applications of Lyndon words are due to Bannai et al. In [11], they employed Lyndon roots of runs to prove the runs conjecture that *the number of runs in a string is bounded by the length of the string*. In the same paper, they presented an algorithm to compute all the runs in a string in linear time that requires the knowledge of all right-maximal Lyndon substrings of the input string with respect to an order of the alphabet and its inverse. The latter result was the major reason for our interest in computing the right-maximal Lyndon substrings of a given string. Though the terms *word* and *string* are interchangeable, and so are the terms *subword* and *substring*, in the following, we prefer to use exclusively the terms *string* and *substring* to avoid confusing the reader.

There are at least two reasons for having an efficient algorithm for identifying all right-maximal Lyndon substrings of a string: firstly, Bannai et al. published in 2017 [11] a linear algorithm to compute all runs in a string that depends on knowing all right-maximal Lyndon substrings of the input string, and secondly, in 2017, Franek et al. in [12] showed a linear equivalence of sorting suffixes and sorting right-maximal Lyndon substrings, inspired by Phase II of a suffix sorting algorithm introduced by Baier in 2015 (Master's thesis [13]) and published in 2016 [14].

The most significant feature of the runs algorithm presented in [11] is that it relies on knowing the right-maximal Lyndon substrings of the input string for some order of the alphabet and for the inverse of that order, while all other linear algorithms for runs rely on Lempel–Ziv factorization of the input string. It also raised the issue about which approach may be more efficient: to compute the Lempel–Ziv factorization or to compute all right-maximal Lyndon substrings. There are several efficient linear algorithms for Lempel–Ziv factorization (e.g., see [15,16] and the references therein).

Interestingly, Kosolobov [17] showed that for a general alphabet, in the decision tree model, the runs problem is easier than the Lempel–Ziv decomposition. His result supports the conjecture that there must be a linear random access memory model algorithm finding all runs.

Baier introduced in [13], and published in [14], a new algorithm for suffix sorting. Though Lyndon strings were never mentioned in [13,14], it was noticed by Cristoph Diegelmann in a personal communication [18] that Phase I of Baier's suffix sort identifies and sorts all right-maximal Lyndon substrings.

The right-maximal Lyndon substrings of a string $x = x[1..n]$ can be best encoded in the so-called *Lyndon array*, introduced in [19] and closely related to the *Lyndon tree* of [20]: an integer array $\mathcal{L}[1..n]$ so that for any $i \in 1..n$, $\mathcal{L}[i] =$ *the length of the right-maximal Lyndon substring starting at the position* i.

In an overview [19], Franek et al. discussed an algorithm based on an iterative application of Duval's Lyndon factorization algorithm [8], which we refer to here as *IDLA*, and an algorithmic scheme based on Hohlweg and Reutenauer's work [20], which we refer to as *SSLA*. The authors were not aware of Baier's algorithm at that time. Two additional algorithms were presented there, a quadratic recursive application of Duval's algorithm and an algorithm *NSV** with possibly $\mathcal{O}(n \log(n))$ worst-case complexity based on ranges that can be compared in constant time for constant alphabets. The correctness of *NSV** and its complexity were discussed there just informally.

The algorithm *IDLA* (see Figure 1) is simple and in-place, so no additional space is required except for the storage for the string and the Lyndon array. It is completely independent of the alphabet of the string and does not require the alphabet to be sorted; all it requires is that the alphabet be ordered, i.e., only pairwise comparisons of the alphabet symbols are needed. Its weakness is its quadratic worst-case complexity, which becomes a problem for longer strings with long right-maximal Lyndon substrings, as one of our experiments showed (see Figure 11 in Section 7).

In our empirical work, we used *IDLA* as a control for comparison and as a verifier of the results. Note that the reason the procedure *MaxLyn* of Figure 1 really computes the longest Lyndon prefix is not obvious and is based on the properties of periods of prefixes; see [8] or Observation 6 and Lemma 11 in [19].

Lemma 1, below, characterizes right-maximal Lyndon substrings in terms of the relationships of the suffixes and follows from the work of Hohlweg and Reutenauer [20]. Though the definition of the proto-Lyndon substring is formally given in Section 2 below, it suffices to say the it means that it is a prefix of a Lyndon substring of the string. The definition of the lexicographic ordering $\prec$ is also given Section 2. The proof of this lemma is delayed to the end of Section 2, where all the technical terms needed are defined.

```
procedure MaxLyn(x[1 .. n], j, Σ, ≺) | integer
    i ← j + 1; max ← 1
    while i ≤ n do
        k ← 0
        while x[j + k] = x[i + k] do
            k ← k + 1
        if x[j + k] ≺ x[i + k] then
            i ← i + k + 1; max ← i − 1
        else
            return max
procedure IDLA(x[1 .. n], j, Σ, ≺) | integer array
    i ← 1
    while i < n do
        L[i] = MaxLyn(x[i .. n], j, Σ, ≺)
        i ← i + 1
    L[n] ← 1
    return L
```

Figure 1. Algorithm *IDLA*.

Lemma 1. *Consider a string $x[1..n]$ over an alphabet ordered by $\prec$.*
A substring $x[i..j]$ is proto-Lyndon if and only if:

(a) $x[i..n] \prec x[k..n]$ *for any* $i < k \leq j$.

A substring $x[i..j]$ is right-maximal Lyndon if and only if:

(b) $x[i..n]$ *is proto-Lyndon and*
(c) *either* $j = n$ *or* $x[j+1..n] \prec x[i..n]$.

Thus, the Lyndon array is an *NSV* (*Next Smaller Value*) array of the inverse suffix array. Consequently, the Lyndon array can be computed by sorting the suffixes, i.e., computing the suffix array, then computing the inverse suffix array, and then applying *NSV* to it; see [19]. Computing the inverse suffix array and applying *NSV* are "naturally" linear, and computing the suffix array can be implemented to be linear; see [19,21] and the references therein. The execution and space characteristics are dominated by those of the first step, i.e., computation of the suffix array. We refer to this scheme as *SSLA*.

In 2018, a linear algorithm to compute the Lyndon array from a given Burrows–Wheeler transform was presented [22]. Since the Burrows–Wheeler transform is computed in linear time from the suffix array, it is yet another scheme of how to obtain the Lyndon array via suffix sorting: compute the suffix array; from the suffix array, compute the Burrows–Wheeler transform, then compute the Lyndon array during the inversion of the Burrows–Wheeler transform. We refer to this scheme as *BWLA*.

The introduction of Baier's suffix sort in 2015 and the consequent realization of the connection to right-maximal Lyndon substrings brought up the realization that there was an elementary (not relying on a pre-processed global data structure such as a suffix array or a Burrows–Wheeler transform) algorithm to compute the Lyndon array, and that, despite its original clumsiness, could be eventually refined to outperform any *SSLA* or *BWLA* implementation: any implementation of a suffix sorting-based scheme requires a full suffix sort and then some additional processing, while Baier's approach is "just" a partial suffix sort; see [23].

In this work, we present two additional algorithms for the Lyndon array not discussed in [19]. The C++ source code of the three implementations *IDLA*, *TRLA*, and *BSLA* is available; see [24]. Note that the procedure *IDLA* is in the `lynarr.hpp` file.

The first algorithm presented here is *TRLA*. *TRLA* is a τ-reduction based Lyndon array algorithm that follows Farach's approach used in his remarkable linear algorithm for suffix tree construction [25] and reproduced very successfully in all linear algorithms for suffix sorting (e.g., see [21,26] and the references therein). Farach's approach follows the Cooley–Tukey algorithm for the fast Fourier transform relying on recursion to lower the quadratic complexity to $O(n \log(n))$ complexity; see [27]. *TRLA* was first introduced by the authors in 2019 (see [28,29]) and presented as a part of Liut's Ph.D. thesis [30].

The second algorithm, *BSLA*, is a Baier's sort-based Lyndon array algorithm. *BSLA* is based on the idea of Phase I of Baier's suffix sort, though our implementation necessarily differs from Baier's. *BSLA* was first introduced at the Prague Stringology Conference 2018 [23] and also presented as a part of Liut's Ph.D. thesis [30] in 2019; here, we present a complete and refined theoretical analysis of the algorithm and a more efficient implementation than that initially introduced.

The paper is structured as follows: In Section 2, the basic notions and terminology are presented. In Section 3, the *TRLA* algorithm is presented and analysed. In Section 4, the *BSLA* algorithm is presented and analysed. In Section 5, the datasets with random strings of various lengths and over various alphabets and other datasets used in the empirical tests are described. In Section 6, the conclusion of the research and the future work are presented. The results of the empirical measurements of the performance of *IDLA*, *TRLA*, and *BSLA* on those datasets are presented in Section 7 in both tabular and graphical forms.

2. Basic Notation and Terminology

Most of the fundamental notions, definitions, facts, and string algorithms can be found in [31–34]. For the ease of access, this section includes those that are directly related to the work herein.

The set of integers is denoted by $\mathbb{Z}$. For two integers $i \leq j$, the *range* $i..j = \{k \in \mathbb{Z} \mid i \leq k \leq j\}$. An *alphabet* is a finite or infinite set of *symbols*, or equivalently called *letters*. We always assume that the sentinel symbol \$ is not in the alphabet and is always assumed to be lexicographically the smallest. A *string* over an alphabet $\mathcal{A}$ is a finite sequence of symbols from $\mathcal{A}$. A *\$-terminated string* over $\mathcal{A}$ is a string over $\mathcal{A}$ terminated by \$. We use the array notation indexing from 1 for strings; thus, $x[1..n]$ indicates a string of length n; the first symbol is the symbol with index 1, i.e., $x[1]$; the second symbol is the symbol with index 2, i.e., $x[2]$, etc. Thus, $x[1..n] = x[1]x[2]...x[n]$. For a \$-terminated string x of length n, $x[n+1] = \$$. The *alphabet of string x*, denoted as $\mathcal{A}_x$, is the set of all distinct alphabet symbols occurring in x.

We use the term *strings over a constant alphabet* if the alphabet is a fixed finite alphabet. The *integer alphabet* is the infinite alphabet $\mathcal{A} = \{0, 1, 2, ...\}$. We use the term *strings over integer alphabet* for the strings over the alphabet $\{0, 1, 2, ...\}$ with an additional constraint that all letters occurring in the string are all smaller than the length of the string, i.e., in this paper, $x[1..n]$ is a string over the integer alphabet if it is a string over the alphabet $\{0, 1, ...n-1\}$. Many authors use a more general definition; for instance, Burkhardt and Kärkkäinen [35] defined it as any set of integers of size $n^{o(1)}$; however, our results can easily be adapted to such more general definitions without changing their essence.

We use a ***bold font*** to denote strings; thus, $\boldsymbol{x}$ denotes a string, while x denotes some other mathematical entity such as an integer. The ***empty string*** is denoted by ε and has length zero. The ***length*** or ***size*** of string $\boldsymbol{x} = \boldsymbol{x}[1..n]$ is n. The length of a string $\boldsymbol{x}$ is denoted by $|\boldsymbol{x}|$. For two strings $\boldsymbol{x} = \boldsymbol{x}[1..n]$ and $\boldsymbol{y} = \boldsymbol{y}[1..m]$, the ***concatenation*** $\boldsymbol{xy}$ is a string $\boldsymbol{u}$

$$\text{where } \boldsymbol{u}[i] = \begin{cases} \boldsymbol{x}[i] \text{ for } i \leq n, \\ \boldsymbol{y}[i-n] \text{ for } n < i \leq n+m. \end{cases}$$

If $\boldsymbol{x} = \boldsymbol{uvw}$, then $\boldsymbol{u}$ is a ***prefix***, $\boldsymbol{v}$ a ***substring***, and $\boldsymbol{w}$ a ***suffix*** of $\boldsymbol{x}$. If $\boldsymbol{u}$ (respectively, $\boldsymbol{v}$, $\boldsymbol{w}$) is empty, then it is called an ***empty prefix*** (respectively, ***empty substring, empty suffix***); if $|\boldsymbol{u}| < |\boldsymbol{x}|$ (respectively, $|\boldsymbol{v}| < |\boldsymbol{x}|$, $|\boldsymbol{w}| < |\boldsymbol{x}|$), then it is called a ***proper prefix*** (respectively, ***proper substring, proper suffix***). If $\boldsymbol{x} = \boldsymbol{uv}$, then $\boldsymbol{vu}$ is called a ***rotation*** or a ***conjugate*** of $\boldsymbol{x}$; if either $\boldsymbol{u} = \varepsilon$ or $\boldsymbol{v} = \varepsilon$, then the rotation is called ***trivial***. A non-empty string $\boldsymbol{x}$ is ***primitive*** if there is no string $\boldsymbol{y}$ and no integer $k \geq 2$ so that $\boldsymbol{x} = \boldsymbol{y}^k = \underbrace{\boldsymbol{yy} \cdots \boldsymbol{y}}_{k \text{ times}}$.

A non-empty string $\boldsymbol{x}$ has a non-trivial border $\boldsymbol{u}$ if $\boldsymbol{u}$ is both a non-empty proper prefix and a non-empty proper suffix of $\boldsymbol{x}$. Thus, both ε and $\boldsymbol{x}$ are trivial borders of $\boldsymbol{x}$. A string without a non-trivial border is call ***unbordered***.

Let $\prec$ be a total order of an alphabet $\mathcal{A}$. The order is extended to all finite strings over the alphabet $\mathcal{A}$: for $\boldsymbol{x} = \boldsymbol{x}[1..n]$ and $\boldsymbol{y} = \boldsymbol{y}[1..n]$, $\boldsymbol{x} \prec \boldsymbol{y}$ if either $\boldsymbol{x}$ is a proper prefix of $\boldsymbol{y}$ or there is a $j \leq \min\{n, m\}$ so that $\boldsymbol{x}[1] = \boldsymbol{y}[1]$, ..., $\boldsymbol{x}[j-1] = \boldsymbol{y}[j-1]$ and $\boldsymbol{x}[j] \prec \boldsymbol{y}[j]$. This total order induced by the order of the alphabet is called a ***lexicographic*** order of all non-empty strings over $\mathcal{A}$. We denote by $\boldsymbol{x} \preceq \boldsymbol{y}$ if either $\boldsymbol{x} \prec \boldsymbol{y}$ or $\boldsymbol{x} = \boldsymbol{y}$. A string $\boldsymbol{x}$ over $\mathcal{A}$ is ***Lyndon*** for a given order $\prec$ of $\mathcal{A}$ if $\boldsymbol{x}$ is strictly lexicographically smaller than any non-trivial rotation of $\boldsymbol{x}$. In particular:

$$\boldsymbol{x} \text{ is Lyndon} \Rightarrow \boldsymbol{x} \text{ is unbordered} \Rightarrow \boldsymbol{x} \text{ is primitive}$$

Note that the reverse implications do not hold: *aba* is primitive but neither unbordered, nor Lyndon, while *acaab* is unbordered, but not Lyndon. A substring $\boldsymbol{x}[i..j]$ of $\boldsymbol{x}[1..n]$, $1 \leq i \leq j \leq n$ is a ***right-maximal Lyndon substring of*** $\boldsymbol{x}$ if it is Lyndon and either $j = n$ or for any $k > j$, $\boldsymbol{x}[i..k]$ is not Lyndon.

A substring $\boldsymbol{x}[i..j]$ of a string $\boldsymbol{x}[1..n]$ is ***proto-Lyndon*** if there is a $j \leq k \leq n$ so that $\boldsymbol{x}[i..k]$ is Lyndon. The ***Lyndon array*** of a string $\boldsymbol{x} = \boldsymbol{x}[1..n]$ is an integer array $\mathcal{L}[1..n]$ so that $\mathcal{L}[i] = j$ where $j \leq n-i$ is a maximal integer such that $\boldsymbol{x}[i..i+j-1]$ is Lyndon. Alternatively, we can define it as an integer array $\mathcal{L}'[1..n]$ so that $\mathcal{L}'[i] = j$ where j is the last position of the right-maximal Lyndon substring starting at the position i. The relationship between those two definitions is straightforward: $\mathcal{L}'[i] = \mathcal{L}[i]+i-1$ or $\mathcal{L}[i] = \mathcal{L}'[i]-i+1$.

Proof of Lemma 1. We first prove the following claim:
Claim. Substring $\boldsymbol{x}[i..j]$ is right-maximal Lyndon if and only if:

(b') $\boldsymbol{x}[i..n] \prec \boldsymbol{x}[k..n]$ for any $i < k \leq j$ and
(c) either $j = n$ or $\boldsymbol{x}[j+1..n] \prec \boldsymbol{x}[i..n]$.

Let $\boldsymbol{x}[i..j]$ be right-maximal Lyndon. We need to show that (b') and (c) hold. Let $i < k \leq j$. Since it is Lyndon, $\boldsymbol{x}[i..j] \prec \boldsymbol{x}[k..j]$. Thus, there is $0 \leq r$ so that $i+r \leq j$ and $j+r \leq j$ and $\boldsymbol{x}[i + \ell] = \boldsymbol{x}[j+\ell]$ for any $0 \leq \ell < r$ and $\boldsymbol{x}[i+r] \prec \boldsymbol{x}[j+r]$. It follows that $\boldsymbol{x}[i..n] \prec \boldsymbol{x}[j..n]$, and (b') is satisfied.

If $j = n$, then (c) holds. Therefore, let us assume that $j < n$.

(1)   If $\boldsymbol{x}[i] \prec \boldsymbol{x}[j+1]$, then $\boldsymbol{x}[i..j+1] \prec \boldsymbol{x}[j+1..j+1]$; together with $\boldsymbol{x}[i..j] \prec \boldsymbol{x}[k..j]$ for any $i < k \leq j$, it shows that $\boldsymbol{x}[i..j+1]$ is Lyndon, contradicting the right-maximality of $\boldsymbol{x}[i..j]$.
(2)   If $\boldsymbol{x}[i] \succ \boldsymbol{x}[j+1]$, then $\boldsymbol{x}[i..n] \succ \boldsymbol{x}[j+1..n]$, and (c) holds.
(3)   If $\boldsymbol{x}[i] = \boldsymbol{x}[j+1]$, then there are a prefix $\boldsymbol{u}$, strings $\boldsymbol{v}$ and $\boldsymbol{w}$, and an integer $r \geq 1$ so that $\boldsymbol{uv} = \boldsymbol{x}[i..j]$, $\boldsymbol{x}[i..n] = \boldsymbol{uvu}^r\boldsymbol{w}$, and $\boldsymbol{x}[j + 1..n] = \boldsymbol{u}^r\boldsymbol{w}$. Let us take a maximal such $\boldsymbol{u}$ and a maximal such r.

(3a) Let $r \geq 2$. Then, since $uv = x[i..j]$ is Lyndon, $u \prec v$, and so, $uu \prec uv$; hence, $u^r w \prec uvu^r w$, and (c) holds.

(3b) Let $r = 1$, i.e., $x[i..n] = uvuw$ and $x[j+1..n] = uw$. There is no common prefix of v and w as it would contradict the maximality of u. There are thus two mutually exclusive cases, either $v[1] \prec w[1]$ or $v[1] \succ w[1]$. Let $v[1] = c_1$ and $w[1] = c_2$. If $c_1 \prec c_2$, then $uc_1 \prec uc_2$, and so, $uv \prec uc_2$. For any $u = u_1 u_2$, $uv \prec u_2 v$ as uv is Lyndon, so $uv \prec u_2 c_2$, giving uvc_2 to be Lyndon, a contradiction with the right-maximality of uv. Therefore, $c_1 \succ c_2$; thus, $uw \prec uvuw$, and (c) holds.

Now, we go in the opposite direction. Assuming (b') and (c), we need to show that $x[i..j]$ is right-maximal Lyndon.

Consider the right-maximal substring of x starting at the position i; it is $x[i..\ell]$ for some $i \leq \ell \leq n$.

(1) If $\ell < j$, then by the first part of this proof, $x[i..n] \succ x[\ell+1..n]$, contradicting the assumption (b') as $\ell+1 \leq j$.

(2) If $j < \ell$, then $j+1 \leq \ell$, and by the first part of this proof, $x[i..n] \prec x[j+1..n]$, contradicting the assumption (c) .

(3) If $\ell = j$, then $x[i..j]$ is right-maximal Lyndon.

Now, we can prove (a). Let $x[i..j]$ be a proto-Lyndon substring of x. By definition, it is a prefix of a Lyndon substring of x and, hence, a prefix of a right-maximal Lyndon substring of x, say $x[i..\ell]$ for some $j \leq \ell \leq n$. It follows from the *claim* that $x[i..n] \prec x[k..n]$ for any $i < k \leq \ell$. Since $j \leq \ell$, (a) holds. For the opposite direction, if (a) holds, then there are two possibilities: either there is $j < \ell \leq n$ so that $x[i..n] \succ x[\ell..n]$ or $x[i..n] \prec x[k..n]$ for any $i < k \leq n$. By the *claim*, in the former case, $x[i..\ell-1]$ is a right-maximal Lyndon substring of x, while in the latter case, $x[i..n]$ is a right-maximal Lyndon substring of x . Thus, in both cases, $x[i..j]$ is a prefix of a Lyndon substring of x.

With (a) proven, we can now replace (b') in the *claim* with (b), completing the proof. $\square$

3. $\mathcal{T}$-Reduction Algorithm (*TRLA*)

The purpose of this section is to introduce a recursive algorithm *TRLA* for computing the Lyndon array of a string. As will be shown below, the most significant aspect is the so-called τ-reduction of a string and how the Lyndon array of the τ-reduced string can be expanded to a partially filled Lyndon array for the whole string, as well as how to compute the missing values. This section thus provides the mathematical justification for the algorithm and, in so doing, proves the correctness of the algorithm. The mathematical understanding of the algorithm provides the bases for the bounding of its worst-case complexity by $O(n \log(n))$ and determining the linearity of the average-case complexity.

The first idea of the algorithm was proposed in Paracha's 2017 Ph.D. thesis [36]. It follows Farach's approach [25]:

(1) reduce the input string x to y;

(2) by recursion, compute the Lyndon array of y; and

(3) from the Lyndon array of y, compute the Lyndon array of x.

The input strings for the algorithm are \$-terminated strings over an integer alphabet. The reduction computed in (1) is important. All linear algorithms for suffix array computations use the proximity property of suffixes: comparing $x[i..n]$ and $x[j..n]$ can be done by comparing $x[i]$ and $x[j]$ and, if they are the same, comparing $x[i+1..n]$ with $x[j+1..n]$. For instance, in the first linear algorithm for the suffix array by Kärkkäinen and Sanders [37], obtaining the sorted suffixes for positions $i \equiv 0 \ (mod \ 3)$ and $i \equiv 1 \ (mod \ 3)$ via the recursive call is sufficient to determine the order of suffixes for the $i \equiv 2 \ (mod \ 3)$ positions, then merging both lists together. However, there is no such proximity property for right-maximal Lyndon substrings, so the reduction itself must have a property that helps determine some of the values of the Lyndon array of x from the Lyndon array of y and computing the rest.

In our algorithm, we use a special reduction, which we call τ-reduction, defined in Section 3.2, that reduces the original string to at most $\frac{1}{2}$ and at least $\frac{2}{3}$ of its length. The algorithm computes y as a τ-reduction of the input string x in Step (1) in linear time. In Step (3), it expands the Lyndon array of the reduced string computed by Step (2) to an incomplete Lyndon array of the original string also in linear time. The incomplete Lyndon array computed in (3) is about $\frac{1}{2}$ to $\frac{2}{3}$ full, and for every position i with an unknown value, the values at positions $i-1$ and $i+1$ are known. In particular, the values at Position 1 and position n are both known. Therefore, much information is provided by the recursive Step (2). For instance, for 00011001, via the recursive call, we would identify the right-maximal Lyndon substrings that are underlined in 00011 001 and would need to compute the missing right-maximal Lyndon substrings that are underlined in 00011001.

However, computing the missing values of the incomplete Lyndon array takes at most $O(n \log(n))$ steps, as we will show, resulting in the overall worst-case complexity of $O(n \log(n))$. When the input string is such that the missing values of the incomplete Lyndon array of the input string can be computed in linear time, the overall execution of the algorithm is linear as well, and thus, the average case complexity will be shown to be linear in the length of the input string.

In the following subsections, we describe the τ-reduction in several steps: first, the τ-pairing, then choosing the τ-alphabet, and finally, the computation of the $\tau(x)$. The τ-reduction may be of some general interest as it preserves (see Lemma 6) some right-maximal Lyndon substrings of the original string.

3.1. τ-Pairing

Consider a \$-terminated string $x = x[1..n]$ whose alphabet $\mathcal{A}_x$ is ordered by $\prec$ where $x[n+1] = \$$ and $\$ \prec a$ for any $a \in \mathcal{A}_x$. A τ-*pair* consists of a pair of adjacent positions from the range $1..n+1$. The τ-pairs are computed by induction:

1. the initial τ-pair is $(1, 2)$;
2. if $(i-1, i)$ is the last τ-pair computed, then:

    if $i = n-1$ **then**
        the next τ-pair is set to $(n, n+1)$
        stop
    elseif $i \geq n$ **then**
        stop
    elseif $x[i-1] \succ x[i]$ **and** $x[i] \preceq x[i+1]$ **then**
        the next τ-pair is set to $(i, i+1)$; repeat 2.
    else
        the next τ-pair is set to $(i+1, i+2)$; repeat 2.

Every position of the input string that occurs in some τ-pair as the first element is labelled **black**; all others are labelled **white**. Note that Position 1 is always black, while the last position n can be either black or white; however, the positions $n-1$ and n cannot be simultaneously both black. Note also that most of the τ-pairs do not overlap; if two τ-pairs overlap, they overlap in a position i such that $1 < i < n$ and $x[i-1] \succ x[i]$ and $x[i] \preceq x[i+1]$. The first position and the last position never figure in an overlap of τ-pairs. Moreover, a τ-pair can be involved in at most one overlap; for an illustration, see Figure 2; for the formal proof see Lemma 2.

Lemma 2. *Let* $(i_1, i_1+1)...(i_k, i_k+1)$ *be the τ-pairs of a string* $x = x[1..n]$. *Then, for any* $j, \ell \in 1..k$:

(1)   *if* $|(i_j, i_j+1) \cap (i_\ell, i_\ell+1)| = 1$, *then for any* $m \neq j, \ell, |(i_j, i_j+1) \cap (i_m, i_m+1)| = 0$,
(2) $|(i_j, i_j+1) \cap (i_\ell, i_\ell+1)| \leq 1$.

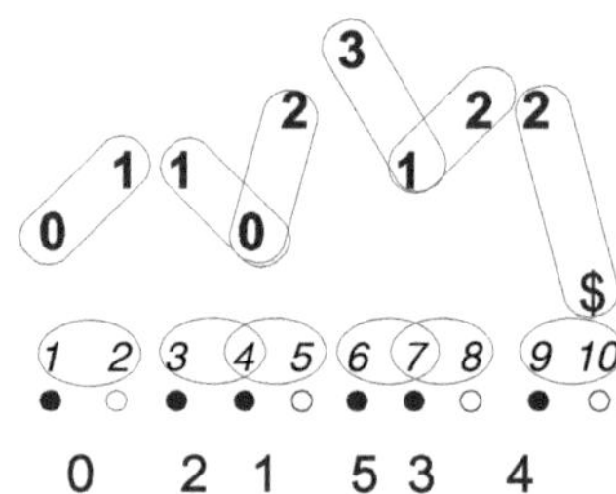

Figure 2. Illustration of the τ-reduction of a string **011023122**. The rounded rectangles indicate symbol τ-pairs; the ovals indicate the τ-pairs. below are the colour labels of the positions; at the bottom is the τ-reduction.

Proof. This is by induction; trivially true for $|x| = 1$ as $(1,2)$ is the only τ-pair. Assume it is true for $|x| \leq n-1$.

1. Case $(i_k, i_k+1) = (n, n+1)$:
   Then, $(i_{k-1}, i_{k-1}+1) = (n-2, n-1)$, and so, $(i_1, i_1+1)...(i_{k-1}, i_{k-1}+1)$ are τ-pairs of $x[1..n-1]$; thus, they satisfy (1) and (2) by the induction hypothesis. However, $(n, n+1) \cap (i_\ell, i_{\ell+1}) = \emptyset$ for $1 \leq \ell < k$, so (1) and (2) hold for $(i_1, i_1+1)...(i_k, i_k+1)$.

2. Cases $(i_k, i_k+1) = (n-1, n)$ and $(i_{k-1}, i_{k-1}+1) = (n-2, n-1)$:
   Therefore, $(i_1, i_1+1)...(i_{k-1}, i_{k-1}+1)$ are τ-pairs of $x[1..n-1]$, and thus, they satisfy (1) and (2) by the induction hypothesis. However, $(i_k, i_k+1) \cap (i_\ell, i_\ell+1) = \emptyset$ for $1 \leq \ell < k-1$, and $(i_k, i_k+1) \cap (i_{k-1}, i_{k-1}+1) = \{i_{k-1}\} = n-1$; so, $|(i_k, i_k+1) \cap (i_{k-1}, i_{k-1}+1)| \leq 1$, and so, (1) and (2) hold for $(i_1, i_1+1)...(i_k, i_k+1)$.

3. Cases $(i_k, i_k+1) = (n-1, n)$ and $(i_{k-1}, i_{k-1}+1) = (n-3, n-2)$:
   Then, $(i_1, i_1+1)...(i_{k-1}, i_{k-1}+1)$ are τ-pairs of $x[1..n-2]$, so they satisfy (1) and (2) by the induction hypothesis. However, $(i_k, i_k+1) \cap (i_\ell, i_\ell+1) = \emptyset$ for $1 \leq \ell < k$, so (1) and (2) hold for $(i_1, i_1+1)...(i_k, i_k+1)$.

$\square$

3.2. τ-Reduction

For each τ-pair $(i, i+1)$, we consider the pair of alphabet symbols $(x[i], x[i+1])$. We call them **symbol τ-pairs**. They are in a total order $\lhd$ induced by $\prec$: $(x[i_j], x[i_j+1]) \lhd (x[i_\ell], x[i_\ell+1])$ if either $x[i_j] \prec x[i_\ell]$, or $x[i_j] = x[i_\ell]$ and $x[i_j+1] \prec x[i_\ell+1]$. They are sorted using the radix sort with a key of size two and assigned letters from a chosen τ-alphabet that is a subset of $\{0, 1, ..., |\tau(x)|\}$ so that the assignment preserves the order. Since the input string is over an integer alphabet, the radix sort is linear.

In the example (Figure 2), the τ-pairs are $(1,2)(3,4)(4,5)(6,7)(7,8)(9,10)$, and so, the symbol τ-pairs are $(0,1)(1,0)(0,2)(3,1)(1,2)(2,\$)$. The sorted symbol τ-pairs are $(0,1)(0,2)(1,0)$ $(1,2)(2,\$)(3,1)$. Thus, we chose as our τ-alphabet $\{0,1,2,3,4,5\}$, and so, the symbol τ-pairs are assigned these letters: $(0,1) \to 0$, $(0,2) \to 1$, $(1,0) \to 2$, $(1,2) \to 3$, $(2,\$) \to 4$, and $(3,1) \to 5$. Note that the assignments respect the order $\lhd$ of the symbols τ-pairs and the natural order $<$ of $\{0,1,2,3,4,5\}$.

The τ-letters are substituted for the symbol τ-pairs, and the resulting string is terminated with $\$$. This string is called the **τ-reduction** of x and denoted $\tau(x)$, and it is a $\$$-terminated string over an integer alphabet. For our running example from Figure 2, $\tau(x) = 021534$. The next lemma justifies calling the above transformation a reduction.

Lemma 3. *For any string x, $\frac{1}{2}|x| \leq |\tau(x)| \leq \frac{2}{3}|x|$.*

Proof. There are two extreme cases; the first is when all the τ-pairs do not overlap at all, then $|\tau(x)| = \frac{1}{2}|x|$; and the second is when all the τ-pairs overlap, then $|\tau(x)| = \frac{2}{3}|x|$. Any other case must be in between. $\square$

Let $\mathcal{B}(x)$ denote the set of all black positions of x. For any $i \in 1..|\tau(x)|$, $b(i) = j$ where j is a black position in x of the τ-pair corresponding to the new symbol in $\tau(x)$ at position i, while $t(j)$ assigns each black position of x the position in $\tau(x)$ where the corresponding new symbol is, i.e., $b(t(j)) = j$ and $t(b(i)) = i$. Thus,

$$1..|\tau(x)| \overset{b}{\underset{t}{\rightleftarrows}} \mathcal{B}(x)$$

In addition, we define p as the mapping of the τ-pairs to the τ-alphabet.

In our running example from Figure 2, $t(1) = 1$, $t(3) = 2$, $t(4) = 3$, $t(6) = 4$, $t(7) = 5$, and $t(9) = 6$, while $b(1) = 1$, $b(2) = 3$, $b(3) = 4$, $b(4) = 6$, $b(5) = 7$, and $b(6) = 9$. For the letter mapping, we get $p(1,2) = 0$, $p(3,4) = 2$, $p(4,5) = 1$, $p(6,7) = 5$, $p(7,8) = 3$, and $p(9,10) = 4$.

3.3. Properties Preserved by τ-Reduction

The most important property of τ-reduction is a preservation of right-maximal Lyndon substrings of x that start at black positions. This means there is a closed formula that gives, for every right-maximal Lyndon substring of $\tau(x)$, a corresponding right-maximal Lyndon substring of x. Moreover, the formula for any black position can be computed in constant time. It is simpler to present the following results using $\mathcal{L}'$, the alternative form of the Lyndon array, the one where the end positions of right-maximal Lyndon substrings are stored rather than their lengths. More formally:

Theorem 2. *Let $x = x[1..n]$; let $\mathcal{L}'_{\tau(x)}[1..m]$ be the Lyndon array of $\tau(x)$; and let $\mathcal{L}'_x[1..n]$ be the Lyndon array of x.*

Then, for any black $i \in 1..n$, $\mathcal{L}'_x[i] = \begin{cases} b(r) & \text{if } b(r) = n \text{ or } x[b(r)+1] \preceq x[i] \\ b(r)+1 & \text{otherwise,} \end{cases}$

where $r = \mathcal{L}'_{\tau(x)}[t(i)]$.

The proof of the theorem requires a series of lemmas that are presented below. First, we show that τ-reduction preserves the relationships of certain suffixes of x.

Lemma 4. *Let $x = x[1..n]$, and let $\tau(x) = \tau(x)[1..m]$. Let $i \neq j$ and $1 \leq i, j \leq n$. If i and j are both black positions, then $x[i..n] \prec x[j..n]$ implies $\tau(x)[t(i)..m] \prec \tau(x)[t(j)..m]$.*

Proof. Since i and j are both black positions, both $t(i)$ and $t(j)$ are defined, and $t(i) \neq t(j)$. Let us assume that $x[i..n] \prec x[j..n]$. The proof is argued in several cases determined by the nature of the relationship $x[i..n] \prec x[j..n]$.

(1)   Case: $x[i..n]$ is a proper prefix of $x[j..n]$.
    Then, $|x[i..n]| = n-i+1 < |x[j..n]| = n-j+1$, and so, $j < i$. It follows that $x[j..j+n-i] = x[i..n]$, and thus, $x[i..n]$ is a border of $x[j..n]$.

    (1a)   Case: $j+n-i$ is black.
        Since n may be either black or white, we need to discuss two cases.

        (1aα)   Case: n is white.
            Since n is white, the last τ-pair of x must be $(n-1, n)$. The τ-pairs of $x[j..j+n-i]$ must be the same as the τ-pairs of $x[i..n]$; the last τ-pair of $x[j..j+n-i]$ must

be $(j+n-i-1, j+n-i)$. Since $j+n-i$ is black by our assumption (1a), the next τ-pair of x must be $(j+n-i, j+n-i+1)$, as indicated in the following diagram:

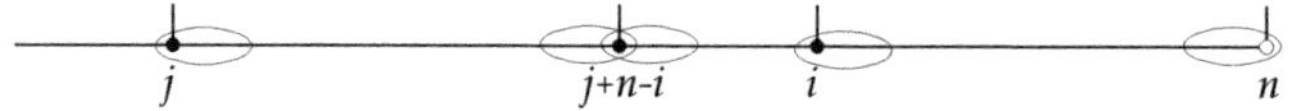

Thus, $\tau(x)[t(j)..t(j+n-i-1)] = \tau(x)[t(i)..t(n-1)]$. Since $t(n-1) = m$, we have $\tau(x)[t(j)..t(j+n-i-1)] = \tau(x)[t(i)..m]$, and so, $\tau(x)[t(i)..m]$ is a proper prefix of $\tau(x)[t(j)..m]$ giving $\tau(x)[t(i)..m] \prec \tau(x)[t(j)..m]$.

(1aβ)   Case: n is black.

Then, the last τ-pair of x must be $(n, n+1)$, and hence, the last τ-pair of $x[j..j+n-i]$, so the next τ-pair is $(j+n-i, j+n-i+1)$; since $n-1$ cannot be black when n is, the situation is as indicated in the following diagram:

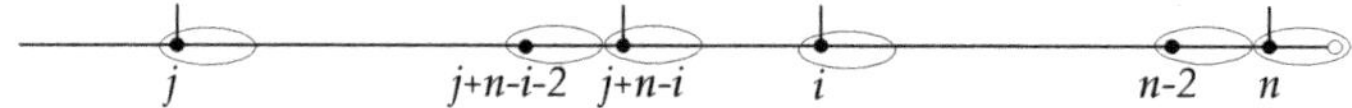

Thus, $\tau(x)[t(i)..t(n-2)] = \tau(x)[t(j)..t(j+n-i-2)]$. Since $x[j+n-i] = x[n]$ and $(x[n], x[n+1]) = (x[n], \$)$, we have $(x[j+n-i], x[j+n-i+1]) \lhd (x[n], x[n+1])$, and so, $\tau(x)[t(j+n-i)] \prec \tau(x)[t(n)]$, giving $\tau(x)[t(j)..t(n)] \prec \tau(x)[t(i)..t(n)]$. Since $t(n) = m$, we have $\tau(x)[t(j)..m] \prec \tau(x)[t(i)..m]$.

(1b)   Case: $j+n-i$ is white.

Then, $j+n-i-1$ is black; hence, $n-1$ is black; so, n must also be white, and thus, $\tau(x)[t(j)..t(j+n-i-1)] = \tau(x)[t(i)..t(n-1)]$, as indicated by the following diagram:

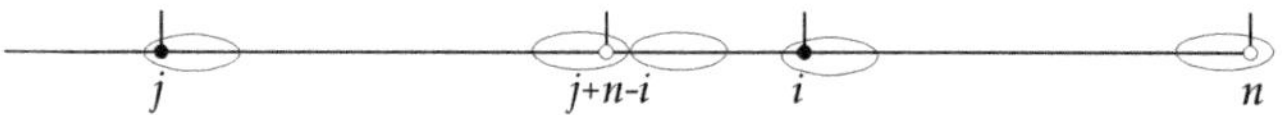

Since $t(n-1) = m$, we have $\tau(x)[t(j)..t(j+n-i-1)] = \tau(x)[t(i)..m]$, and so, $\tau(x)[t(i)..m]$ is a proper prefix of $\tau(x)[t(j)..m]$, giving $\tau(x)[t(i)..m] \prec \tau(x)[t(j)..m]$.

(2)   Case: $x[i] \prec x[j]$ or $(x[i] = x[j]$ and $x[i+1] \prec x[j+1])$.

Then, $(x[i], x[i+1]) \lhd (x[j], x[j+1])$, and so, $\tau(x)[t(i)] \prec \tau(x)[t(j)]$, and thus, $\tau(x)[t(i)..m] \prec \tau(x)[t(j)..m]$.

(3)   Case: for some $\ell \geq 3$, $x[i..i+\ell-1] = x[j..j+\ell-1]$, while $x[i+\ell] \prec x[j+\ell]$.

First note that $i+\ell-2$ and $j+\ell-2$ are either both black, or both are white:

- If $i+\ell-2$ is white, then the τ-pairs $((i, i+1), ..., (i+\ell-3, i+\ell-2))$ of $x[i..n]$ correspond one-to-one to the τ-pairs $((j, j+1), ..., (j+\ell-3, j+\ell-2))$ of $x[j..n]$. To determine what follows $(i+\ell-3, i+\ell-2)$, we need to know the relationship between the values $x[i+\ell-3]$, $x[i+\ell-2]$, and $x[i+\ell-1]$. Since $x[i+\ell-3] = x[j+\ell-3]$, $x[i+\ell-2] = x[j+\ell-2]$, and $x[i+\ell-1] = x[j+\ell-1]$, the values $x[j+\ell-3]$, $x[j+\ell-2]$, and $x[j+\ell-1]$ have the same relationship, and thus, the τ-pair following $(j+\ell-3, j+\ell-2)$ will be the "same" as the τ-pair following $(i+\ell-3, i+\ell-2)$. Since $i+\ell-2$ is white, the τ-pair following $(i+\ell-3, i+\ell-2)$ is $(i+\ell-1, i+\ell)$, and so, the τ-pair following $(j+\ell-3, j+\ell-2)$ is $(j+\ell-1, j+\ell)$, making $j+\ell-2$ white as well.
- If $i+\ell-2$ is black, then the τ-pairs $((i, i+1), ..., (i+\ell-2, i+\ell-1))$ of $x[i..n]$ correspond one-to-one to the τ-pairs $((j, j+1), ..., (j+\ell-2, j+\ell-1))$ of $x[j..n]$. It follows that $j+\ell-2$ is black as well.

We proceed by discussing these two cases for the colours of $i+\ell-2$ and $j+\ell-2$.

(3a)   Case when $i+\ell-2$ and $j+\ell-2$ are both white.

Therefore, we have the τ-pairs $((i, i+1), ..., (i+\ell-3, i+\ell-2), (i+\ell-1, i+\ell))$ for $x[i..n]$

that correspond one-to-one to the τ-pairs $\big((j, j+1), ..., (j+\ell-3, j+\ell-2), (j+\ell-1, j+\ell)\big)$ for $x[j..n]$. It follows that $\tau(x)[t(i)..t(i+\ell-3)] = \tau(x)[t(j)..t(j+\ell-3)]$. $\tau(x)[t(i+\ell-1)] = p(i+\ell-1, i+\ell)$ and $\tau(x)[t(j+\ell-1)] = p(j+\ell-1, j+\ell)$. Since $x[i+\ell-1] = x[j+\ell-1]$ and, by our assumption (3), $x[i+\ell] \prec x[j+\ell]$, it follows that $(i+\ell-1, i+\ell) \lhd (j+\ell-1, j+\ell)$, giving $p(i+\ell-1, i+\ell) \prec p(j+\ell-1, j+\ell)$, and so, $\tau(x)[t(i+\ell-1)] \prec \tau(x)[t(j+\ell-1)]$. Since $t(i+\ell-3)+1 = t(i+\ell-1)$ and $t(j+\ell-3)+1 = t(j+\ell-1)$, we have $\tau(x)[t(i)..m] \prec \tau(x)[t(j)..m]$.

(3b)   Case when $i+\ell-2$ and $j+\ell-2$ are both black.

Therefore, we have the τ-pairs $\big((i, i+1), ..., (i+\ell-2, i+\ell-1)\big)$ for $x[i..n]$ that correspond one-to-one to the τ-pairs $\big((j, j+1), ..., (j+\ell-2, j+\ell-1)\big)$ for $x[j..n]$. It follows that $\tau(x)[t(i)..t(i+\ell-2)] = \tau(x)[t(j)..t(j+\ell-2)]$.

We need to discuss the four cases based on the colours of $i+\ell-1$ and $j+\ell-1$.

(3bα)   Both $i+\ell-1$ and $j+\ell-1$ are black.

It follows that the next τ-pair for $x[i..n]$ is $(i+\ell-1, i+\ell)$, and the next τ-pair for $x[j..n]$ is $(j+\ell-1, j+\ell)$. It follows that $t(i+\ell-2)+1 = t(i+\ell-1)$ and $t(j+\ell-2)+1 = t(j+\ell-1)$. Hence, $\tau(x)[t(i+\ell-2)+1] = p(i+\ell-1, i+\ell)$ and $\tau(x)[t(j+\ell-2)+1] = p(j+\ell-1, j+\ell)$. Since $x[i+\ell-1] = x[j+\ell-1]$ and, by Assumption (3), $x[i+\ell] \prec x[j+\ell]$, we have $(x[i+\ell-1], x[i+\ell]) \lhd (x[j+\ell-1], x[j+\ell])$, and so, $p(x[i+\ell-1], x[i+\ell]) \prec p(x[j+\ell-1], x[j+\ell])$, giving us $\tau(x)[t(i+\ell-2)+1] \prec \tau(x)[t(j+\ell-2)+1]$. It follows that $\tau(x)[t(i)..m] \prec \tau(x)[t(j)..m]$.

(3bβ) $i+\ell-1$ is white, and $j+\ell-1$ is black.

It follows that the next τ-pair for $x[i..n]$ is $(i+\ell, i+\ell+1)$, and the next τ-pair for $x[j..n]$ is $(j+\ell-1, j+\ell)$. It follows that $t(i+\ell-2)+1 = t(i+\ell)$, while $t(j+\ell-2)+1 = t(j+\ell-1)$. Thus, $\tau(x)[t(i+\ell-2)+1] = p(i+\ell, i+\ell+1)$ and $\tau(x)[t(j+\ell-2)+1] = p(j+\ell-1, j+\ell)$. Since $j+\ell-1$ is black, we know that $x[j-\ell-2] \succ x[j-\ell-1] \preceq x[j+\ell]$. Since $x[i+\ell-2] = x[j+\ell-2]$ and $x[i+\ell-1] = x[j+\ell-1]$, we have $x[i-\ell-2] \succ x[i-\ell-1]$, and so, $x[i-\ell-1] \succ x[i+\ell]$, as otherwise, $i-\ell-1$ would be black. This gives us $x[j+\ell-1] \succ x[i+\ell]$. Thus, $(x[i+\ell], x[i+\ell+1]) \lhd (x[j+\ell-1], x[j+\ell])$, giving $p(i+\ell, i+\ell+1) \prec p(j+\ell-1, j+\ell)$ and, ultimately, $\tau(x)[t(i+\ell)] \prec \tau(x)[t(j+\ell-1)]$. The last step is to realize that $t(i+\ell-2)+1 = t(i+\ell)$ and $t(j+\ell-2)+1 = t(j+\ell-2)$, which gives us $\tau(x)[t(i+\ell-2)+1] \prec \tau(x)[t(j+\ell-2)+1]$. It follows that $\tau(x)[t(i)..m] \prec \tau(x)[t(j)..m]$.

(3bγ) $i+\ell-1$ is black, and $j+\ell-1$ is white.

It follows that the next τ-pair for $x[i..n]$ is $(i+\ell-1, i+\ell)$, and the next τ-pair for $x[j..n]$ is $(j+\ell, j+\ell+1)$. It follows that $t(i+\ell-2)+1 = t(i+\ell-1)$, while $t(j+\ell-2)+1 = t(j+\ell)$. Thus, $\tau(x)[t(i+\ell-2)+1] = p(i+\ell-1, i+\ell)$ and $\tau(x)[t(j+\ell-2)+1] = p(j+\ell, j+\ell+1)$. Since $i+\ell-1$ is black, we know that $x[i+\ell-2] \succ x[i+\ell-1] \preceq x[i+\ell] \prec x[j+\ell]$, where the last inequality is our Assumption (3). Therefore, $x[j+\ell-1] = x[i+\ell-1] \prec x[j+\ell]$. Thus, $(x[i+\ell-1], x[i+\ell]) \lhd (x[j+\ell], x[j+\ell+1])$, giving $p(i+\ell-1, i+\ell) \prec p(j+\ell, j+\ell+1)$, $\tau(x)[t(i+\ell-1)] \prec \tau(x)[t(j+\ell)]$, and ultimately, $\tau(x)[t(i+\ell-2)+1] = \tau(x)[t(i+\ell-1)] \prec \tau(x)[t(j+\ell)] = \tau(x)[t(j+\ell-2)+1]$. It follows that $\tau(x)[t(i)..m] \prec \tau(x)[t(j)..m]$.

(3bδ)   Both $i+\ell-1$ and $j+\ell-1$ are white.

Then, the next τ-pair for $x[i..n]$ is $(i+\ell, i+\ell+1)$, and the next τ-pair for $x[j..n]$ is $(j+\ell, j+\ell+1)$. It follows that $t(i+\ell-2)+1 = t(i+\ell)$, while $t(j+\ell-2)+1 = t(j+\ell)$. Thus, $\tau(x)[t(i+\ell-2)+1] = p(i+\ell, i+\ell+1)$ and

$$\tau(x)[t(j+\ell-2)+1] = p(j+\ell, j+\ell+1).$$ Since $x[i+\ell-1] = x[j+\ell-1]$ and, by our Assumption (3), $x[i+\ell] \prec x[j+\ell]$, $(x[i+\ell], x[i+\ell-1]) \prec (x[j+\ell], x[j+\ell-1])$, giving $p(i+\ell, i+\ell+1) \prec p(j+\ell, j+\ell+1)$, $\tau(x)[t(i+\ell)] \prec \tau(x)[t(j+\ell)]$, and ultimately, $\tau(x)[t(i+\ell-2)+1] = \tau(x)[t(i+\ell)] \prec \tau(x)[t(j+\ell)] = \tau(x)[t(j+\ell-2)+1]$. It follows that $\tau(x)[t(i)..m] \prec \tau(x)[t(j)..m]$.

$\square$

Lemma 5 shows that τ-reduction preserves the proto-Lyndon property of certain proto-Lyndon substrings of x.

Lemma 5. *Let $x = x[1..n]$, and let $\tau(x) = \tau(x)[1..m]$. Let $1 \le i < j \le n$. Let $x[i..j]$ be a proto-Lyndon substring of x, and let i be a black position.*

Then, $\begin{cases} \tau(x)[t(i)..t(j)] \text{ is proto-Lyndon} & \text{if } j \text{ is black} \\ \tau(x)[t(i)..t(j-1)] \text{ is proto-Lyndon} & \text{if } j \text{ is white.} \end{cases}$

Proof. Let us first assume that j is black.
Since both i and j are black, $t(i)$ and $t(j)$ are defined. Let $i_1 = t(i)$, $j_1 = t(j)$, and consider k_1, so that $i_1 < k_1 \le j_1$. Let $k = b(k_1)$. Then, $t(k) = k_1$ and $i < k \le j$, and so, $x[i..n] \prec x[k..n]$ by Lemma 1 as $x[i..j]$ is proto-Lyndon. It follows that $\tau(x)[t(i)..m] \prec \tau(x)[t(k)..m]$ by Lemma 4. Thus, $\tau(x)[i_1..m] \prec \tau(x)[k_1..m]$ for any $i_1 < k_1 \le j_1$, and so, $\tau(x)[i_1..j_1]$ is proto-Lyndon by Lemma 1.

Now, let us assume that j is white.
Then, $j-1$ is black, and $x[i..j-1]$ is proto-Lyndon, so as in the previous case, $\tau(x)[t(i)..t(j-1)]$ is proto-Lyndon. $\square$

Now, we can show that τ-reduction preserves some right-maximal Lyndon substrings.

Lemma 6. *Let $x = x[1..n]$, and let $\tau(x) = \tau(x)[1..m]$. Let $1 \le i < j \le n$. Let $x[i..j]$ be a right-maximal Lyndon substring, and let i be a black position.*

Then, $\begin{cases} \tau(x)[t(i)..t(j)] \text{ is a right-maximal Lyndon substring} & \text{if } j \text{ is black} \\ \tau(x)[t(i)..t(j-1)] \text{ is a right-maximal Lyndon substring} & \text{if } j \text{ is white.} \end{cases}$

Proof. Since $x[i..j]$ is Lyndon and hence proto-Lyndon, by Lemma 5, we know that $\tau(x)[t(i)..t(j)]$ is proto-Lyndon for j black, while for white j, $\tau(x)[t(i)..t(j-1)]$ is proto-Lyndon. Thus, in order to conclude that the respective strings are right-maximal Lyndon substrings, we only need to prove that the property (c) of Lemma 1 holds in both cases.

Since $x[i..j]$ is right-maximal Lyndon, either $j = n$ or $x[j+1..n] \prec x[i..n]$ by Lemma 1, giving $j = n$ or $x[j+1] \preceq x[i]$. Since $x[i..j]$ is Lyndon and hence unbordered, $x[i] \prec x[j]$. Thus, either $j = n$ or $x[j+1] \preceq x[i] \prec x[j]$.

If $j = n$, then there are two simple cases. If n is white, $n-1$ is black and $m = t(n-1)$, so $t(j-1) = m$, giving us (c) of Lemma 1 for $\tau(x)[t(i)..t(j-1)]$. On the other hand, if n is black, then $m = t(n)$, and so, $m = t(j)$ giving us (c) of Lemma 1 for $\tau(x)[t(i)..t(j)]$.

Thus, in the following, we can assume that $j < n$ and that $x[j+1] \preceq x[i] \prec x[j]$. We will proceed by discussing two possible cases, one where j is black and the other where j is white.

(1) Case: j is black.
We need to show that either $t(j) = m$ or $\tau(x)[t(i)..m] \succ \tau(x)[t(j)+1..m]$.
If $j = n$, then $t(j) = m$, and we are done. Thus, we can assume that $j < n$. We must show that $\tau(x)[t(j)+1..m] \prec \tau(x)[t(i)..m]$.

(1a) Case: $x[j+1] \preceq x[j+2]$.
Then, $x[j] \succ x[j+1]$ and $x[j+1] \preceq x[j+2]$, and so, $j+1$ is black. It follows that $t(j) + 1 = t(j+1)$. By Lemma 4, $\tau(x)[t(j+1)..m] \prec \tau(x)[t(i)..m]$ because $x[j+1..n] \prec x[i..n]$, thus $\tau(x)[t(j)+1..m] \prec \tau(x)[t(i)..m]$.

(1b) Case: $x[j+1] \succ x[j+2]$.
Then, $x[j] \succ x[i] \succeq x[j+1] \succ x[j+2]$. It follows that the τ-pair $(j, j+1)$ is followed by a τ-pair $(j+2, j+3)$, and thus, $t(j)+1 = t(j+2)$. Thus, $(x[j+2], x[j+3]) \lhd (x[i], x[i+1]) \lhd (x[j], x[j+1])$; hence, $p(j+2, j+3) \prec p(i, i+1) \prec p(j, j+1)$. Since $\tau(x)[t(j)+1] = p(j+2, j+3)$, $\tau(x)[t(i)] = p(j, i+1)$, and $\tau(x)[t(j)] = p(j, j+1)$, it follows that $\tau(x)[t(j)+1] \prec \tau(x)[t(i)]$, and so, $\tau(x)[t(j)+1..m] \prec \tau(x)[t(i)..m]$.

(2) Case: j is white.
We need to prove that $\tau(x)[t(j-1)+1..m] \prec \tau(x)[t(i)..m]$. Since j is white, necessarily both $j-1$ and $j+1$ are black and $t(j-1)+1 = t(j+1)$. By Lemma 5, $\tau(x)[t(i)..t(j-1)]$ is proto-Lyndon as both i and $j-1$ are black and $x[i..j-1]$ is proto-Lyndon. Since $x[i..n] \succ x[j+1..n]$ and both i and $j+1$ are black, by Lemma 4, we get $\tau(x)[t(i)..m] \succ \tau(x)[t(j+1)..m] = \tau(x)[t(j-1)+1..m]$.

$\square$

Now, we are ready to tackle the proof of Theorem 2.

Proof of Theorem 2. Let $\mathcal{L}'_x[i] = j$ where i is black. Then, $t(i)$ is defined, and $x[i..j]$ is a right-maximal Lyndon substring of x. We proceed by analysis of the two possible cases of the label for the position j. Let $(*)$ denote the condition from the theorem, i.e.,

$$(*) \qquad b\big(\mathcal{L}'_{\tau(x)}[t(i)]\big) = n \quad or \quad x\big[b\big(\mathcal{L}'_{\tau(x)}[t(i)]\big)+1\big] \preceq x[i]$$

(1) Case: j is black.

Then, by Lemma 6, $\tau(x)[t(i)..t(j)]$ is a right-maximal Lyndon substring of $\tau(x)$; hence, $\mathcal{L}'_{\tau(x)}[t(i)] = t(j)$. Therefore, $b\big(\mathcal{L}'_{\tau(x)}[t(i)]\big) = b(t(j)) = j = \mathcal{L}'_x[i]$. We have to also prove that the condition $(*)$ holds.
If $j = n$, then the condition $(*)$ holds. Therefore, assume that $j < n$. Since $x[i..j]$ is right-maximal, by Lemma 1, $x[j+1..n] \prec x[i..n]$, and so, $x[j+1] \preceq x[i]$. Then, $x\big[b\big(\mathcal{L}'_{\tau(x)}[t(i)]\big)+1\big] = x[b(t(j))+1] = x[j+1] \preceq x[i]$.

(2) Case: j is white.

Then, $j-1$ is black, and $\tau(x)[t(j-1)] = p(j-1, j)$. By Lemma 6, $\tau(x)[t(i)..t(j-1)]$ is a right-maximal Lyndon substring of $\tau(x)$; hence, $\mathcal{L}'_{\tau(x)}[t(i)] = t(j-1)$, so $b\big(\mathcal{L}'_{\tau(x)}[t(i)]\big) = b(t(j-1)) = j-1$, giving $b\big(\mathcal{L}'_{\tau(x)}[t(i)]+1\big) = j$.
We want to show that the condition $(*)$ does not hold.
If $b\big(\mathcal{L}'_{\tau(x)}[t(i)]\big) = n$, then $j-1 = n$, which is impossible as $j \leq n$. Since $x[i..j]$ is Lyndon, $x[i] \prec x[j]$, and so, $x[i] \prec x\big[b\big(\mathcal{L}'_{\tau(x)}[t(i)]+1\big)\big]$. Thus, Condition $(*)$ does not hold.

$\square$

3.4. Computing $\mathcal{L}'_x$ from $\mathcal{L}'_{\tau(x)}$

Theorem 2 indicates how to compute the partial $\mathcal{L}'_x$ from $\mathcal{L}'_{\tau(x)}$. The procedure is given in Figure 3.

$$
\begin{aligned}
&\textbf{for } i \leftarrow 1 \textbf{ to } n \\
&\quad \textbf{if } i = 1 \textbf{ or } \big(x[i-1] \succ x[i] \textbf{ and } x[i] \preceq x[i+1]\big) \textbf{ then} \\
&\quad\quad \textbf{if } x\big[b\big(\mathcal{L}'_{\tau(x)}[t(i)]\big)+1\big] \preceq x[i] \textbf{ then} \\
&\quad\quad\quad \mathcal{L}'_x[i] \leftarrow b\big(\mathcal{L}'_{\tau(x)}[t(i)]\big) \\
&\quad\quad \textbf{else} \\
&\quad\quad\quad \mathcal{L}'_x[i] \leftarrow b\big(\mathcal{L}'_{\tau(x)}[t(i)]\big)+1 \\
&\quad \textbf{else} \\
&\quad\quad \mathcal{L}'_x[i] \leftarrow nil
\end{aligned}
$$

Figure 3. Computing the partial Lyndon array of the input string.

To compute the missing values, the partial array is processed from right to left. When a missing value at position i is encountered (note that it is recognized by $\mathcal{L}'_x[i] = nil$), the Lyndon array $\mathcal{L}'_x[i+1..n]$ is completely filled, and also, $\mathcal{L}'_x[i-1]$ is known. Recall that $\mathcal{L}'_x[i+1]$ is the ending position of the right-maximal Lyndon substring starting at the position $i+1$. In several cases, we can determine the value of $\mathcal{L}'_x[i]$ in constant time:

1. if $i = n$, then $\mathcal{L}'_x[i] = i$.
2. if $x[i] \succ x[i+1]$, then $\mathcal{L}'_x[i] = i$.
3. if $x[i] = x[i+1]$ and $\mathcal{L}'_x[i+1] = i+1$ and either $i+1 = n$ or $i+1 = \mathcal{L}'_x[i-1]$, then $\mathcal{L}'_x[i] = i$.
4. if $x[i] \prec x[i+1]$ and $\mathcal{L}'_x[i+1] = i+1$ and either $i+1 = n$ or $i+1 = \mathcal{L}'_x[i-1]$, then $\mathcal{L}'_x[i] = i+1$.
5. if $x[i] \preceq x[i+1]$ and $\mathcal{L}'_x[i+1] > i+1$ and either $\mathcal{L}'_x[i+1] = n$ or $\mathcal{L}'_x[i+1] = \mathcal{L}'_x[i-1]$, then $\mathcal{L}'_x[i] = \mathcal{L}'_x[i+1]$.

We call such points *easy*. All others will be referred to as *hard*. For a *hard* point i, it means that $x[i]$ is followed by at least two consecutive right-maximal Lyndon substrings before reaching either $\mathcal{L}'_x[i-1]$ or n, and we might need to traverse them all.

The **while** loop, seen in Figure 4's procedure, is the likely cause of the $\mathcal{O}(n\,\log(n))$ complexity. At first glance, it may seem that the complexity might be $\mathcal{O}(n^2)$; however, the doubling of the length of the string when a hard point is introduced actually trims it down to an $\mathcal{O}(n\,\log(n))$ worst-case complexity. See Section 3.5 for more details and Section 7 for the measurements and graphs.

$$
\begin{aligned}
&\mathcal{L}'_x[n] \leftarrow n \\
&\textbf{for } i \leftarrow n-1 \textbf{ downto } 2 \\
&\quad \textbf{if } \mathcal{L}'[i] = nil \textbf{ then} \\
&\quad\quad \textbf{if } x[i] \succ x[i+1] \textbf{ then} \\
&\quad\quad\quad \mathcal{L}'[i] \leftarrow i \\
&\quad\quad \textbf{else} \\
&\quad\quad\quad \textbf{if } \mathcal{L}'[i-1] = i-1 \textbf{ then} \\
&\quad\quad\quad\quad stop \leftarrow n \\
&\quad\quad\quad \textbf{else} \\
&\quad\quad\quad\quad stop \leftarrow \mathcal{L}'[i-1] \\
&\quad\quad\quad \mathcal{L}'[i] \leftarrow \mathcal{L}'[i+1] \\
&\quad\quad\quad \textbf{while } \mathcal{L}'[i] < stop \textbf{ do} \\
&\quad\quad\quad\quad \textbf{if } x[i..\mathcal{L}'[i]] \prec x[\mathcal{L}'[i]+1..\mathcal{L}'[\mathcal{L}'[i]+1]] \textbf{ then} \\
&\quad\quad\quad\quad\quad \mathcal{L}'[i] \leftarrow \mathcal{L}'[\mathcal{L}'[i]+1] \\
&\quad\quad\quad\quad \textbf{else} \\
&\quad\quad\quad\quad\quad \textbf{break}
\end{aligned}
$$

Figure 4. Computing missing values of the Lyndon array of the input string.

Consider our running example from Figure 2. Since $\tau(x) = 021534$, we have $\mathcal{L}'_{\tau(x)}[1..6] = 6, 2, 6, 4, 6, 6$ giving $\mathcal{L}'_x[1..9] = 9, \bullet, 3, 9, \bullet, 6, 9, \bullet, 9$. Computing $\mathcal{L}'_x[8]$ is easy as $x[8] = x[9]$, and so, $\mathcal{L}'_x[8] = 8$. $\mathcal{L}'_x[5]$ is more complicated and an example of a hard point: we can extend the right-maximal Lyndon substring from $\mathcal{L}'_x[6]$ to the left to 23, but no more, so $\mathcal{L}'_x[5] = 6$. Computing $\mathcal{L}'_x[2]$ is again easy as $x[2] = x[3]$, and so, $\mathcal{L}'_x[2] = 2$. Thus, $\mathcal{L}'_x[1..9] = 9, 2, 3, 9, 6, 6, 9, 8, 9$.

3.5. The Complexity of TRLA

To determine the complexity of the algorithm, we attach to each position i a counter $red[i]$ initialized to zero. Imagine a *hard* point j indicated by the following diagram:

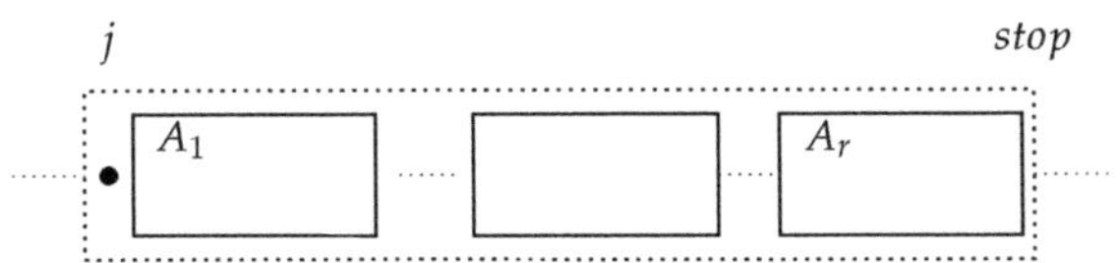

A_1 represents the right-maximal Lyndon substring starting at the position $j+1$; A_2 represents the right-maximal Lyndon substring following immediately A_1 and so forth. To make j a *hard* point, $r \geq 2$ and $x[j] \preceq x[j+1]$. The value of *stop* is determined by:

$$stop = \begin{cases} \mathcal{L}'_x[j-1] & \textit{if } \mathcal{L}'_x[j-1] > j-1 \\ n & \textit{otherwise} \end{cases}.$$

To determine the right-maximal Lyndon substring starting at the *hard* position j, we need first to check if A_1 can be left-extended by $x[j]$ to make jA_1 Lyndon; we are using abbreviated notation jA_1 for the substring $x[j..k]$ where $A_1 = x[j+1..k]$; in simple words, jA_1 represents the left-extension of A_1 by one position. If jA_1 is proto-Lyndon, we have to check whether A_2 can be left-extended by jA_1 to a Lyndon substring. If jA_1A_2 is Lyndon, we must continue until we check whether $jA_1A_2...A_{r-1}$ is Lyndon. If so, we must check whether $jA_1...A_r$ is Lyndon. We need not go beyond *stop*.

How do we check if $jA_1...A_k$ can left-extend A_{k+1} to a Lyndon substring? If $jA_1...A_k \succeq A_{k+1}$, we can stop, and $jA_1...A_k$ is the right-maximal Lyndon substring starting at position j. If $jA_1...A_k \prec A_{k+1}$, we need to continue. Since *stop* is the last position of the right-maximal Lyndon substring at the position $j-1$ or n, we are assured to stop there. When comparing the substring $jA_1...A_k$ with A_{k+1}, we increment the counter $red[i]$ at every position of A_{k+1} used in the comparison. When done with the whole array, the value of $red[i]$ represents how many times i was used in various comparisons, for any position i.

Consider a position i that was used k times for $k \geq 4$, i.e., $red[i] = k$. In the next four diagrams and related text, the upper indices of A and C do not represent powers; they are just indices. The next diagram indicates the configuration when the counter $red[i]$ was incremented for the first time in the comparison of $j_1 A_1^1...A_{r_1-1}^1$ and $A_{r_1}^1$ during the computation of the missing value $\mathcal{L}'_x[j_1]$ where:

$$stop_1 = \begin{cases} \mathcal{L}'_x[j_1-1] & \textit{if } \mathcal{L}'_x[j_1-1] > j_1-1 \\ n & \textit{otherwise} \end{cases}$$

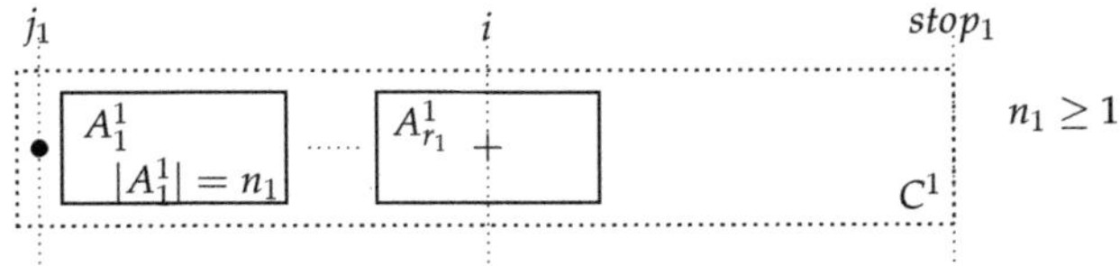

The next diagram indicates the configuration when the counter $red[i]$ was incremented for the second time in the comparison of $j_2 A_1^2 ... A_{r_2-1}^2$ and $A_{r_2}^2$ during the computation of the missing value $\mathcal{L}'_x[j_2]$ where:

$$stop_2 = \begin{cases} \mathcal{L}'_x[j_2-1] & if \ \mathcal{L}'_x[j_2-1] > j_2-1 \\ n & otherwise \end{cases}$$

The next diagram indicates the configuration when the counter $red[i]$ was incremented for the third time in the comparison of $j_3 A_1^3 ... A_{r_3-1}^3$ and $A_{r_3}^3$ during the computation of the missing value $\mathcal{L}'_x[j_3]$ where:

$$stop_3 = \begin{cases} \mathcal{L}'_x[j_3-1] & if \ \mathcal{L}'_x[j_3-1] > j_3-1 \\ n & otherwise \end{cases}$$

The next diagram indicates the configuration when the counter $red[i]$ was incremented for the fourth time in the comparison of $j_4 A_1^4 ... A_{r_4-1}^4$ and $A_{r_4}^4$ during the computation of the missing value $\mathcal{L}'_x[j_4]$ where:

$$stop_4 = \begin{cases} \mathcal{L}'_x[j_4-1] & if \ \mathcal{L}'_x[j_4-1] > j_4-1 \\ n & otherwise \end{cases}$$

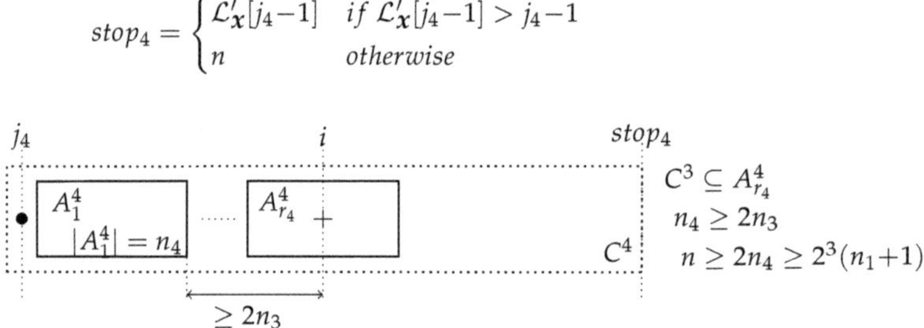

and so forth until the k-th increment of $red[i]$. Thus, if $red[i] = k$, then $n \geq 2^{k-1}(n_1+1) \geq 2^k$ as $n_1+1 \geq 2$. Thus, $n \geq 2^k$, and so, $k \leq \log(n)$. Thus, either $k < 4$ or $k \leq \log(n)$. Therefore, the overall complexity is $\mathcal{O}(n \ \log(n))$.

To show that the average case complexity is linear, we first recall that the overall complexity of *TRLA* is determined by the procedure filling the missing values. We showed above that there are at most $\log(n)$ missing values (*hard* positions) that cannot be determined in constant time. We overestimate the number of strings of length n over an alphabet of size Σ, $2 \leq \Sigma \leq n$, which will force a non-linear computation, by assuming that every possible $\log(n)$ subset of indices with any possible letter assignment forces the worst performance. Thus, there are $\Sigma^n - \binom{n}{\log(n)}\Sigma^{\log(n)}$ strings that are processed in linear time, say with a constant K_1, and there are $\binom{n}{\log(n)}\Sigma^{\log(n)}$ strings that are processed in the worst time, with a constant K_2. Let $K = \max(K_1, K_2)$. Then, the average time is bounded by:

$$\frac{\left(\Sigma^n - \binom{n}{\log(n)}\Sigma^{\log(n)}\right)Kn + \binom{n}{\log(n)}\Sigma^{\log(n)}Kn\log(n)}{\Sigma^n} =$$

$$Kn + Kn\frac{\binom{n}{\log(n)}\Sigma^{\log(n)}}{\Sigma^n}\left(\log(n) - 1\right) \le$$

$$Kn + Kn\frac{\binom{n}{\log(n)}\Sigma^{\log(n)}}{\Sigma^n}\log(n) \le$$

$$Kn + Kn\frac{n^{\log(n)}\Sigma^{\log(n)}}{\log(n)!\Sigma^n}\log(n) \le$$

$$Kn + Kn\frac{n^{2\log(n)}}{2^n} \le$$

$$Kn + Kn = 2Kn$$

for $n \ge 2^7$. The last step follows from the fact that $n^{2\log(n)} \le 2^n$ for any $n \ge 2^7$.

The combinatorics of the processing is too complicated to ascertain whether the worst-case complexity is linear or not. We tried to generate strings that might give the worst performance. We used three different formulas to generate the strings, nesting the white indices that might require non-constant computation: the dataset `extreme_trla` of binary strings is created using the recursive formula $u_{k+1} = 00u_k0u_k$, using the first 100 shortest binary Lyndon strings as the start u_0. The moment the size u_k exceeds the required length of the string, the recursion stops, and the string is trimmed to the required length. For the `extreme_trla1` dataset, we used the same approach with the formula $u_{k+1} = 000u_k00u_k$, and for the `extreme_trla2` dataset, we used the formula $u_{k+1} = 0000u_k00u_k$.

The space complexity of our C++ implementation is bounded by $9n$ integers. This upper bound is derived from the fact that a Tau object (see `Tau.hpp` [24]) requires $3n$ integers of space for a string of length n. Therefore, the first call to *TRLA* requires $3n$, the next recursive call at most $3\frac{2}{3}n$, the next recursive call at most $3(\frac{2}{3})^2n$, ...; thus, $3n + 3\frac{2}{3}n + 3(\frac{2}{3})^2n + 3(\frac{2}{3})^3n + ... = 3n(1 + \frac{2}{3} + (\frac{2}{3})^2 + (\frac{2}{3})^3 + (\frac{2}{3})^4 + ...) = 3n\frac{1}{1-\frac{2}{3}} = 9n$. However, it should be possible to bring it down to $6n$ integers.

4. The Algorithm *BSLA*

The purpose of this section is to present a linear algorithm *BSLA* for computing the Lyndon array of a string over an integer alphabet. The algorithm is based on a series of refinements of a list of groups of indices of the input string. The refinement is driven by a group that is already complete, and the refinement process makes the immediately preceding group also complete. In turn, this newly completed group is used as the driver of the next round of the refinement. In this fashion, the refinement proceeds from right to left until all the groups in the list are complete. The initial list of groups consists of the groups of indices with the same alphabet symbol. The section contains proper definitions of all these terms—group, complete group, and refinement. In the process of refinement, each newly created group is assigned a specific substring of the input string referred to as the context of the group. Throughout the process, the list of the groups is maintained in an increasing lexicographic order by their contexts. Moreover, at every stage, the contexts of all the groups are Lyndon substrings of x with an additional property that the contexts of the complete groups are right-maximal Lyndon substrings. Hence, when the refinement is completed, the contexts of all the groups in the list represent all the right-maximal Lyndon substrings of x. The mathematics of the process of refinement is necessary in order to ascertain its correctness and completeness and to determine the worst-case complexity of the algorithm.

4.1. Notation and Basic Notions of BSLA

For the sake of simplicity, we fix a string $x = x[1..n]$ for the whole Section 4.1; all the definitions and the observations apply and refer to this x.

A ***group*** G is a non-empty set of indices of x. The group G is assigned a ***context***, i.e., a substring $con(G)$ of x with the property that for any $i \in G$, $x[i..i+|con(G)|-1] = con(G)$. If $i \in G$, then $\mathcal{C}(i)$ denotes the occurrence of the context of G at the position i, i.e., the substring $\mathcal{C}(i) = x[i..i+|con(G)|-1]$. We say that a group G' is ***smaller than*** or ***precedes*** a group G'' if $con(G') \prec con(G'')$.

Definition 1. *An ordered list of groups* $\langle G_k, G_{k-1}, ..., G_2, G_1 \rangle$ *is a* ***group configuration*** *if:*

(C$_1$) $G_k \cup G_{k-1} \cup ... \cup G_2 \cup G_1 = 1..n;$
(C$_2$) $G_j \cap G_\ell = \emptyset$ *for any* $1 \leq \ell < j \leq k;$
(C$_3$) $con(G_k) \prec con(G_{k-1}) \prec ... \prec con(G_2) \prec con(G_1);$
(C$_4$) *For any* $j \in 1..k$, $con(G_j)$ *is a Lyndon substring of* x.

Note that (C$_1$) and (C$_2$) guarantee that $\langle G_k, G_{k-1}, ..., G_2, G_1 \rangle$ is a disjoint partitioning of $1..n$. For $i \in \{1,..,n\}$, $gr(i)$ denotes the unique group to which i belongs, i.e., if $i \in G_t$, then $gr(i) = G_t$. Note that using this notation, $\mathcal{C}(i) = x[i..i+|con(gr(i))|-1]$.

The mapping ***prev*** is defined by $prev(i) = \max\{j < i \mid con(gr(j)) \prec con(gr(i))\}$ if such j exists, otherwise $prev(i) = nil$.

For a group G from a group configuration, we define an equivalence $\sim$ on G as follows: $i \sim j$ iff $gr(prev(i)) = gr(prev(j))$ or $prev(i) = prev(j) = nil$. The symbol $[i]_\sim$ denotes the class of equivalence $\sim$ that contains i, i.e., $[i]_\sim = \{j \in G \mid j \sim i\}$. If $prev(i) = nil$, then the class $[i]_\sim$ is called trivial. An interesting observation states that if G is viewed as an ordered set of indices, then a non-trivial $[i]_\sim$ is an interval:

Observation 7. *Let G be a group from a group configuration for x. Consider an $i \in G$ such that $prev(i) \neq nil$. Let $j_1 = \min[i]_\sim$ and $j_2 = \max[i]_\sim$. Then, $[i]_\sim = \{j \in G \mid j_1 \leq j \leq j_2\}$.*

Proof. Since $prev(j_1)$ is a candidate to be $prev(j)$, $prev(j) \neq nil$ and $prev(j_1) \leq prev(j) \leq prev(j_2) = prev(j_1)$, so $prev(j) = prev(j_1) = prev(j_2)$. $\square$

On each non-trivial class of $\sim$, we define a relation $\approx$ as follows: $i \approx j$ iff $|j-i| = |con(G)|$; *in simple terms, it means that the occurrence $\mathcal{C}(i)$ of $con(G)$ is immediately followed by the occurrence $\mathcal{C}(j)$ of $con(G)$. The transitive closure of $\approx$ is a relation of equivalence, which we also denote by $\approx$. The symbol* $[i]_\approx$ *denotes the class of equivalence $\approx$ containing i, i.e., $[i]_\approx = \{j \in [i]_\sim \mid j \approx i\}$.*

For each j from a non-trivial $[i]_\sim$, we define the ***valence*** by $val(j) = |[i]_\approx|$. *In simple terms, $val(i)$ is the number of elements from $[i]_\sim$ that are $\approx i$. Thus, $1 \leq val(i) \leq |G|$.*

Interestingly, if G is viewed as an ordered set of indices, then $[i]_\approx$ is a subinterval of the interval $[i]_\sim$:

Observation 8. *Let G be a group from a group configuration for x. Consider an $i \in G$ such that $prev(i) \neq nil$. Let $j_1 = \min[i]_\approx$ and $j_2 = \max[i]_\approx$. Then, $[i]_\approx = \{j \in [i]_\sim \mid j_1 \leq j \leq j_2\}$.*

Proof. We argue by contradiction. Assume that there is an $j \in [i]_\sim$ so that $j_1 < j < j_2$ and $j \notin [i]_\approx$. Take the minimal such j. Consider $j' = j - |con(G)|$. Then, $j' \in [i]_\sim$, and since, $j' < j$, $j' \in [i]_\approx$ due to the minimality of j. Therefore, $i \approx j' \approx j$, and so, $j \approx i$, a contradiction. $\square$

Definition 2. *A group G is **complete** if for any $i \in G$, the occurrence $C(i)$ of $con(G)$ is a right-maximal Lyndon substring of x.*
*A group configuration $\langle G_k, G_{k-1}, ..., G_2, G_1 \rangle$ is **t-complete**, $1 \leq t \leq k$, if*

(C_5) *the groups $G_t, ..., G_1$ are complete;*

(C_6) *the mapping prev is **proper** on G_t:*
   for any $i \in G_t$, if $prev(i) \neq nil$ and $v = val(i)$, then there are $i_1, ..., i_v \in G_t$, $i \in \{i_1, ..., i_v\}$, $prev(i) = prev(i_1) = ... = prev(i_v)$, and so that $C(prev(i))C(i_1)...C(i_v)$ is a prefix of $x[j..n]$;

(C_7) *the family $\{C(i) \mid i \in 1..n\}$ is **proper**:*
   (a) if $C(j)$ is a proper substring of $C(i)$, i.e., $C(j) \subsetneq C(i)$, then $con(G_t) \prec con(gr(j))$,
   (b) if $C(i)$ is followed immediately by $C(j)$, i.e., when $i + |con(gr(i))| = j$, and $C(i) \prec C(j)$, then $con(gr(j)) \preceq con(G_t)$;

(C_8) *the family $\{C(i) \mid i \in 1..n\}$ has the **Monge** property, i.e., if $C(i) \cap C(j) \neq \emptyset$, then $C(i) \subseteq C(j)$ or $C(j) \subseteq C(i)$.*

The condition (C_6) is all-important for carrying out the refinement process (see (R_3) below). The conditions (C_7) and (C_8) are necessary for asserting that the condition (C_6) is preserved during the refinement process.

4.2. The Refinement

For the sake of simplicity, we fix a string $x = x[1..n]$ for the whole Section 4.2; all the definitions, lemmas, and theorems apply and refer to this x.

Lemma 9. *Let $\mathcal{A}_x = \{a_1, ..., a_k\}$ and $a_1 \prec a_2 \prec ... \prec a_k$. For $1 \leq \ell \leq k$, define $G_\ell = \{i \in 1..n \mid x[i] = a_{k+1-\ell}\}$ with context $a_{k+1-\ell}$. Then, $\langle G_k, ..., G_1 \rangle$ is a one-complete group configuration.*

Proof. (C_1), (C_2), (C_3), and (C_4) are straightforward to verify. To verify (C_5), we need to show that G_1 is complete. Any occurrence of a_k in x is a right-maximal Lyndon substring, so G_1 is complete.
To verify (C_6), consider $j = prev(i)$ and $val(i) = v$ for $i \in G_1$. Consider any r such that $j < r < i$. If $x[r] \neq a_k$, then $prev(i) < r$, which contradicts the definition of $prev$. Hence, $x[r] = a_k$, and so, $x[j+1] = ... = x[i] = ...x[j+v+1] = a_k$, while $x[j] = a_q$ for some $q < k$. It follows that $x[j..n]$ has $a_q(a_k)^v$ as a prefix.
The condition $(C_7(a))$ is trivially satisfied as no $C(i)$ can have a proper substring. If $C(i)$ is immediately followed by $C(j)$ and $C(i) \prec C(j)$, then $C(i) = x[i]$, $j = i+1$, $C(j) = x[i+1]$, and $x[i] \prec x[i+1]$. Then, $con(C(j)) = x[i+1] \preceq a_k = con(G_1)$, so $(C_7(b))$ is also satisfied.
To verify (C_8), consider $C(i) \cap C(j) \neq \emptyset$. Then, $C(i) = x[i] = x[j] = C(j)$. $\square$

Let $\langle G_k, ..., G_t, ..., G_1 \rangle$ by a t-complete group configuration. The refinement is driven by the group G_t, and it might only partition the groups that precede it, i.e., the groups $G_k, ..., G_{t+1}$, while the groups $G_t, ..., G_1$ remain unchanged.

(R_1) Partition G_t into classes of the equivalence $\sim$.
   $G_t = [i_1]_\sim \cup [i_2]_\sim \cup ... \cup [i_p]_\sim \cup X$ where $X = \{i \in G_t \mid prev(i) = nil\}$ may be possibly empty and $i_1 < i_2 < ... < i_p$.

(R_2) Partition every class $[i_\ell]_\sim$, $1 \leq \ell \leq p$, into classes of the equivalence $\approx$.
   $[i_\ell]_\sim = [j_{\ell,1}]_\approx \cup [j_{\ell,2}]_\approx \cup ... \cup [j_{\ell,m_\ell}]_\approx$ where $val(j_{\ell,1}) < val(j_{\ell,2}) < ... < val(j_{\ell,m_\ell})$.

(R_3) Therefore, we have a list of classes in this order: $[j_{1,1}]_\approx$, $[j_{1,2}]_\approx$, $\cdots$ $[j_{1,m_1}]_\approx$, $[j_{2,1}]_\approx$, $[j_{2,2}]_\approx$, $\cdots$ $[j_{2,m_2}]_\approx$, $\cdots$, $[j_{p,1}]_\approx$, $[j_{p,2}]_\approx$, $\cdots$ $[j_{p,m_p}]_\approx$. This list is processed from left to right. Note that for each $i \in [j_{\ell,k}]_\approx$, $prev(i) \in gr(j_{\ell,k})$, and $val(i) = val(j_{\ell,k})$.
   For each $j_{\ell,k}$, move all elements $\{prev(i) \mid i \in [j_{\ell,k}]_\approx\}$ from the group $gr(prev(j_{\ell,k}))$ into a new

group H, place H in the list of groups right after the group $gr(prev(j_{\ell,k}))$, and set its context to $con(gr(prev(j_{\ell,k})))con(gr(j_{\ell,k}))^{val(j_{\ell,k})}$. (*Note, that this "doubling of the contexts" is possible due to* (C_6)). Then, update *prev*:

> All values of *prev* are correct except possibly the values of *prev* for indices from H. It may be the case that for $i \in H$, there is $i' \in gr(j_{\ell,k})$, so that $prev(i) < i'$, so $prev(i)$ must be reset to the maximal such i'. (*Note that before the removal of H from $gr(j_{\ell,k})$, the index i' was not eligible to be considered for $prev(i)$ as i and i' were both from the same group.*)

Theorem 3 shows that having a t-complete group configuration $\langle G_k, ..., G_{t+1}, G_t, ..., G_1 \rangle$ and refining it by G_t, then the resulting system of groups is a $(t+1)$-complete group configuration. This allows carrying out the refinement in an iterative fashion.

Theorem 3. *Let $Conf = \langle G_k, ..., G_{t+1}, G_t, ..., G_1 \rangle$ be a t-complete group configuration, $1 \le t$. After performing the refinement of Conf by group G_t, the resulting system of groups denoted as $Conf'$ is a $(t+1)$-complete group configuration.*

Proof. We carry the proof in a series of claims. The symbols $gr()$, $con()$, $\mathcal{C}()$, $prev()$, and $val()$ denote the functions for $Conf$, while $gr'()$, $con'()$, $\mathcal{C}'()$, $prev'()$, and $val'()$ denote the functions for $Conf'$. When a group G_{t+1} is partitioned, a part of it is moved as the next group in the list, and we call it H_{t+1}; thus, $G_{t+1} \prec H_{t+1} \prec G_t$. For details, please see (R_3) above.

Claim 1. *$Conf'$ is a group configuration, i.e., (C_1), (C_2), (C_3), and (C_4) for $Conf'$ hold.*

Proof of Claim 1. (C_1) and (C_2) follow from the fact that the process is a refinement, i.e., a group is either preserved as is or is partitioned into two or more groups. The doubling of the contexts in Step (R_3) guarantees that the increasing order of the contexts is preserved, i.e., (C_3) holds. For any $j \in G_t$ so that $j = prev(i) \ne nil$, $con(gr(prev(j)))$ is Lyndon, and $con(gr(j))$ is also Lyndon, while $con(gr(prev(j))) \prec con(gr(j))$, so $con(gr(prev(j)))con(gr(j))^{val(j)}$ is Lyndon as well; thus, (C_4) holds.
To illustrate the concatenation: let us call $con(gr(prev(j)))$ as A and $con(gr(j))$ as B, and let $val(j) = m$, then we know that A is Lyndon and B is Lyndon and $A \prec B$; so, AB^m is clearly Lyndon as if A and B were letters. $\square$

This concludes the proof of Claim 1.

Claim 2. *$\{\mathcal{C}'(i) \mid i \in 1..n\}$ is proper and has the Monge property, i.e., (C_7) and (C_8) for $Conf'$ hold.*

Proof of Claim 2. Consider $\mathcal{C}'(i)$ for some $i \in 1..n$. There are two possibilities:

1. $\mathcal{C}'(i) = \mathcal{C}(i)$ or
2. $\mathcal{C}'(i) = \mathcal{C}(i)\mathcal{C}(i_1)...\mathcal{C}(i_v)$, for some $i_1, i_2, ..., i_v \in G_t$, so that for any $1 \le \ell \le v$, $i = prev(i_\ell)$, $\mathcal{C}(i_\ell) = con(G_t)$, $v = val(i_\ell)$ and for any $1 \le \ell < k$ and $i_{\ell+1} = i_\ell + |con(G_t)|$. Note that $con(gr(i)) \prec con(G_t)$.

Consider $\mathcal{C}'(i)$ and $\mathcal{C}'(j)$ for some $1 \le i < j \le n$.

1. Case $\mathcal{C}'(i) = \mathcal{C}(i)$ and $\mathcal{C}'(j) = \mathcal{C}(j)$.

    (a) Show that $(C_7(a))$ holds.
    If $\mathcal{C}'(j) \subsetneqq \mathcal{C}'(i)$, then $\mathcal{C}(j) \subsetneqq \mathcal{C}(i)$, and so, by $(C_7(a))$ for $Conf$, $con(G_t) \prec con(gr(j))$, and thus, $con'(H_{t+1}) \prec con(G_t) \prec con(gr(j)) = con'(gr'(j))$. Therefore, $(C_7(a))$ for $Conf'$ holds.

   (b)   Show that (C_8) holds. If $C'(i) \cap C'(j) \neq \varnothing$, then $C(i) \cap C(j) \neq \varnothing$, so $C(j) \subseteq C(i)$, and so, $C'(j) \subseteq C'(i)$; so, (C_8) for *Conf'* holds.

2.   Case $C'(i) = C(i)$ and $C'(j) = C(j)C(j_1)..C(j_w)$, where $w = val(j_1)$, $C(j_1) = ... = C(j_w) = con(G_t)$, and $j_1 \approx ... \approx j_w$.

   (a)   Show that $(C_7(a))$ holds.
If $C'(j) \subsetneq C'(i)$, then $C(j)C(j_1)..C(j_w) \subsetneq C(i)$; hence, $C(j) \subsetneq C(i)$, and so, by $(C_7(a))$ for *Conf*, $con(G_t) \prec con(gr(j))$. By the t-completeness of *Conf*, $C(j)$ is a right-maximal Lyndon substring, a contradiction with $C(j)C(j_1)..,C(j_w)$ being Lyndon. This is an impossible case.

   (b)   Show that (C_8) holds.
If $C'(i) \cap C'(j) \neq \varnothing$, then $C(j) \subseteq C(i)$ by (C_8) for *Conf*. By $(C_7(a))$ for *Conf*, $C(j)$ cannot be a suffix of $C(i)$ as $con(gr(j)) \prec con(G_t)$. Hence, $C(i) \cap C(j_1) \neq \varnothing$, and so, $C(j)C(j_1) \subseteq C(i)$; and since $C(j_1)$ cannot be a suffix of $C(i)$ as $gr(j_1) = G_t$, it follows that $C(i) \cap C(j_2) \neq \varnothing$, ..., ultimately giving $C(j)C(j_1)...C(j_w) \subseteq C(i)$. Therefore, (C_8) for *Conf'* holds.

3.   Case $C'(i) = C(i)C(i_1)..C(i_v)$ and $C'(j) = C(j)$, where $v = val(i_1)$, $C(i_1) = ... = C(i_v) = con(G_t)$, and $i_1 \approx ... \approx i_v$.

   (a)   Show that $(C_7(a))$ holds.
If $C'(j) \subsetneq C'(i)$, then either $C(j) \subsetneq C(i)$, which implies by $(C_7(a))$ for *Conf* that $con(G_t) \prec con(gr(j))$, giving $con'(H_{t+1}) \prec con'(G_t) = con(G_t) \prec con(gr(j)) = con'(gr'(j))$, or $C(j) \subseteq C(i_\ell)$ for some $1 \leq \ell \leq v$. If $C(j) = C(i_\ell)$, then $gr(j) = gr(i_\ell) = G_t$, giving $con'(H_{t+1}) \prec con(G_t) = con(gr(j))$. Therefore $(C_7(a))$ for *Conf'* holds.

   (b)   Show that (C_8) holds.
Let $C'(i) \cap C'(j) \neq \varnothing$. Consider $\mathcal{D} = \{i_\ell \mid 1 \leq \ell \leq v \text{ and } C(j) \cap C(i_\ell) \neq \varnothing\}$.
Assume that $\mathcal{D} \neq \varnothing$:

By (C_8) for *Conf*, either $C(j) \subseteq \bigcup_{i_\ell \in \mathcal{D}} C(i_\ell) \subseteq C'(i)$, and we are done, or $\bigcup_{i_\ell \in \mathcal{D}} C(i_\ell) \subseteq C(j)$. Let i_k be the smallest element of $\mathcal{D}$. Since $C(i_k)$ cannot be the prefix of $C(j)$, it means that $i_k = i_1$. Since $C(i_1)$ cannot be a prefix of $C(j)$, it means that $C(i) \cap C(j) \neq \varnothing$, and so, $C(j) \subseteq C(i)$, which contradicts the fact that $C(j) \subseteq \bigcup_{i_\ell \in \mathcal{D}} C(i_\ell) \subseteq C'(i)$.

Assume that $\mathcal{D} = \varnothing$:

Then, $C(i) \cap C(j) \neq \varnothing$, and so, by (C_8) for *Conf*, $C(j) \subseteq C(i) \subseteq C'(i)$ as $i < j$.

4.   Case $C'(i) = C(i)C(i_1)..C(i_v)$ and $C'(j) = C(j)C(j_1)...C(j_w)$, where $v = val(i_1)$, $C(i_1) = ... = C(i_v) = con(G_t)$, and $i_1 \approx ... \approx i_v$ and where $v = val(j_1)$, $C(j_1) = ... = C(j_w) = con(G_t)$, and $j_1 \approx ... \approx j_w$.

   (a)   Show that $(C_7(a))$ holds.
Let $C'(j) \subsetneq C'(i)$. Then, either $C(j) \subseteq C(i)$, and so, $con(G_t) \prec con(gr(j))$, implying that $C(j)$ is maximal contradicting $C(j)C(j_1)...C(j_w)$ being Lyndon. Thus, $C(j) \subsetneq C(i_\ell)$ for some $1 \leq \ell \leq v$. However, then, $con(G_t) \prec con(gr(j))$, implying that $C(j)$ is maximal, again a contradiction. This is an impossible case.

   (b)   Show that (C_8) holds.
Let $C'(i) \cap C'(j) \neq \varnothing$. Let us first assume that $C(i) \cap C(j) \neq \varnothing$. Then, $C(j) \subseteq C(i)$. Since $C(j)$ cannot be a suffix of $C(i)$, it follows that $C(i) \cap C(j_1) \neq \varnothing$. Therefore, $C(j)C(j_1) \subseteq C(i)$. Repeating this argument leads to $C(j)C(j_1)...C(j_w) \subseteq C(i)$, and we are done.
Therefore, assume that $C(i) \cap C(j) = \varnothing$. Let $1 \leq \ell \leq v$ be the smallest such that $C(i_\ell) \cap C(j) \neq \varnothing$. Such an ℓ must exist. Then, $i_\ell \leq j$. If $i_\ell = j$, then either $C(i_\ell)$ is a prefix of $C(j)$ or vice versa, both impossibilities; hence, $i_\ell < j$. Repeating the same arguments as for i, we get that $C(j)C(j_1)..C(j_w) \subseteq C(i_\ell)$, and so, we are done.

It remains to show that $(C_7(b))$ for *Conf'* holds.

Consider $C'(i)$ immediately followed by $C'(j)$ with $C'(i) \prec C'(j)$.

1. Assume that $gr'(j) \in \{G_{t-1}, ..., G_1\}$.

   Then, $con(G_t) = con'(G_t)$, $gr(j) = gr'(j)$, and $con(gr(j)) = con'(gr'(j))$. If $C'(i) = C(i)$, then $C(i) \prec C(j)$, and $C(i)$ is immediately followed by $C(j)$, so by $(C_7(b))$ for *Conf*, we have a contradiction. Thus, $C'(i) = C(i)C(i_1)...C(i_v)$ for $v = val(i)$ and $con(gr(i_v)) = con(G_t) \prec con(gr(j))$, and $C(i_v)$ is immediately followed by $C(j)$, a contradiction by $(C_7(b))$ for *Conf*.

2. Assume that $gr'(j) = G_t$.

   Then, the group $gr(i)$ is partitioned when refining by G_t, and so, $C'(i) = con'(gr'(i)) = con(gr(i))C(j)^v$ for $v = val(j)$. Since $C'(i)$ is immediately followed by $C'(j) = con(G_t)$, we have again a contradiction, as it implies that $val(j) = v+1$.

$\square$

This concludes the proof of Claim 2.

Claim 3. *The function prev' is proper on* H_{t+1}, *i.e.,* (C_6) *for Conf' holds.*

Proof of Claim 3. Let $j = prev'(i)$ and $i \in H_{t+1}$ with $val'(i) = v$. Then, $|[i]_{\approx}| = v$, and so, $[i]_{\approx} = \{i_1, ..., i_v\}$, where $i_1 < i_2 < ... < i_v$. Hence, $i_1, ..., i_v \in H_{t+1}$, $C'(i_1) = ... = C'(i_v) = con'(H_{t+1})$, and $j = prev'(i) = prev'(i_1) = ... = prev'(i_v)$, and so, $j < i_1$. It remains to show that $C'(j)C'(i_1)...C'(i_v)$ is a prefix of $x[j..n]$. It suffices to show that $C'(j)$ is immediately followed by $C'(i_1)$.

If $C'(j) \cap C'(i_1) \neq \emptyset$, then by the Monge property (C_8), $C'(i_1) \subseteq C'(j)$ as $j < i_1$, and so, by $(C_7(a))$, $con'(H_{t+1}) \prec con'(gr'(i_1)) = con'(H_{t+1})$, a contradiction.

Thus, $C'(j) \cap C'(i_1) = \emptyset$. Set $j_1 = j+|con'(gr'(j))|$. It follows that $j_1 \leq i_1$. Assume that $j_1 < i_1$. Since $j = prev'(i_1)$ and $j < i_1$, $con'(gr'(j_1)) \succeq con'(gr'(i_1)) = con'(H_{t+1})$. Since $j_1 \notin H_{t+1}$, $con'(gr'(j_1)) \succ con'(H_{t+1})$. Consider $C'(j_1)$. If $C'(j_1) \cap C'(i_1) \neq \emptyset$, then by (C_8), $C'(i_1) \subseteq C'(j_1)$, and so, by $(C_7(a))$, $con'(H_{t+1}) \prec con'(gr'(i_1)) = con'(H_{t+1})$, a contradiction. Thus, $C'(j_1) \cap C'(i_1) = \emptyset$. Since $C'(j_1)$ immediately follows $C'(j)$, by $(C_7(b))$, $con'(gr'(j_1)) \preceq con'(H_{t+1})$, a contradiction. Therefore, $j_1 = i_1$, and so, $prev'$ is proper on H_{t+1}. $\square$

This concludes the proof of Claim 3.

Claim 4. H_{t+1} *is a complete group, i.e.,* (C_5) *for Conf' holds.*

Proof of Claim 4. Assume that there is $i \in H_{t+1}$ so that $C'(i)$ is not maximal, i.e., for some $k \geq i+|con'(H_{t+1})|$, $x[i..k]$ is a right-maximal Lyndon substring of x.

Either $k = n$ and so $con'(gr'(k)) = x[k]$, and so, $C'(k)$ is a suffix of $x[i..k]$, or $k < n$, and then, $x[k+1] \prec x[k]$, since $x[k+1] \preceq x[k]$ implies that $x[i..k+1]$ is Lyndon, a contradiction with the right-maximality of $x[i..k]$. Consider $C'(k)$, then $C'(k) \subseteq x[i..k]$, and so, $C'(k) = x[k]$.

Therefore, there is j_1 so that $i+|con'(H_{t+1})| \leq j_1 \leq k$, and $C'(j_1)$ is a suffix of $x[i..k]$. Take the smallest such j_1. If $j_1 = i+|con'(H_{t+1})|$, then $C'(i) \prec C'(j_1)$ as $x[i..k] = C'(i)C'(j_1)$ is Lyndon. By $(C_7(b))$, $C'(j_1) \preceq con'(H_{t+1})$, so we have $con'(H_{t+1}) = C'(i) \prec C'(j_1) \preceq con'(H_{t+1})$, a contradiction.

Therefore, $j_1 > i+|con'(H_{t+1})|$. Consider $x[j_1-1]$. If $x[j_1-1] \preceq x[j_1]$, $x[j_1-1..k]$ is Lyndon, and since $x[j_1..k] = C'(j_1)$, $x[j_1-1..k]$ would be a context of $gr'(j_1-1)$, this contradicts the fact j_1 was chosen to be the smallest such one. Therefore, $x[j_1-1] \succ x[j_1]$, and so, $con'(gr'(j_1-1)) = x[j_1-1]$. Thus, there is j_2, $i+|con'(H_{t+1})| \leq j_2 < j_1 \leq k$, and $C'(j_2)$ is a suffix of $x[i..j_1-1]$. Take the smallest such j_2. If $C'(j_2) \prec C'(j_1)$, then by $(C_7(b))$, $C'(j_1) \preceq con'(H_{t+1})$, a contradiction. Hence, $C'(j_2) \succeq C'(j_1)$. If $j_2 = i+i+|con'(H_{t+1})|$, then $x[i..k] = C'(i)C'(j_2)C'(j_1)$, and so, by $(C_7(b))$, $C'(j_2) \preceq con'(H_{t+1})$, a contradiction. Hence, $i+|con'(H_{t+1})| < j_2$.

The same argument done for j_2 can now be done for j_3. We end up with $i+|con'(H_{t+1})| \leq j_3 < j_2 < j_1 \leq k$ and with $C'(j_3) \succeq C'(j_2) \succeq C'(j_1) \succ con'(H_{t+1})$. If $i+|con'(H_{t+1})| = j_3$, then we have a contradiction, so $i + |con'(H_{t+1})| < j_3$. These arguments can be repeated only finitely many times, and we obtain $i + |con'(H_{t+1})| = j_\ell < j_{\ell-1} < ... < j_2 < j_1 \leq k$ so that $x[i..k] = C'(i)C'(j_\ell)C'(j_{\ell-1}...C'(j_2)C'(j_1)$, which is a contradiction.

Therefore, our initial assumption that $C'(i)$ is not maximal always leads to a contradiction. $\square$

This concludes the proof of Claim 4.

The four claims show that all the conditions (C_1) ... (C_8) are satisfied for *Conf'*, and that proves Theorem 3. $\square$

As the last step, we show that when the process of refinement is completed, all right-maximal Lyndon substrings of x are identified and sorted via the contexts of the groups of the final configuration.

Theorem 4.
Let $Conf_1 = \langle G_{k_1}^1, G_{k_1-1}^1, ..., G_2^1, G_1^1 \rangle$ *with* $gr_1()$, $con_1()$, $C_1()$, $prev_1()$, *and* $val_1()$ *be the initial 1-complete group configuration from Lemma 9.*

Let $Conf_2 = \langle G_{k_2}^2, G_{k_2-1}^2, ..., G_2^2, G_1^2 \rangle$ *with* $gr_2()$, $con_2()$, $C_2()$, $prev_2()$, *and* $val_2()$ *be the 2-complete group configuration obtained from* $Conf_1$ *through the refinement by the group* G_1^1.

Let $Conf_3 = \langle G_{k_3}^3, G_{k_3-1}^3, ..., G_2^3, G_1^3 \rangle$ *with* $gr_3()$, $con_3()$, $C_3()$, $prev_3()$, *and* $val_3()$ *be the 3-complete group configuration obtained from* $Conf_2$ *through the refinement by the group* G_2^2.

...

Let $Conf_r = \langle G_{k_r}^r, G_{k_r-1}^r, ..., G_2^r, G_1^r \rangle$ *with* $gr_r()$, $con_r()$, $C_r()$, $prev_r()$, *and* $val_r()$ *be the r-complete group configuration obtained from* $Conf_{r-1}$ *through the refinement by the group* G_{r-1}^{r-1}. *Let* $Conf_r$ *be the final configuration after the refinement runs out.*

Then, $x[i..k]$ *is a right-maximal Lyndon substring of* x *iff* $x[i..k] = C_r(i) = con_r(gr_r(i))$.

Proof. That all the groups of *Conf_r* are complete follows from Theorem 3, and hence, every $C_r(i)$ is a right-maximal Lyndon string. Let $x[i..k]$ be a right-maximal Lyndon substring of x. Consider $C_r(i)$; since it is maximal, it must be equal to $x[i..k]$. $\square$

4.3. Motivation for the Refinement

The process of refinement is in fact a process of the gradual revealing of the Lyndon substrings, which we call the ***water draining method***:

(a) lower the water level by one;
(b) extend the existing Lyndon substrings; *the revealed letters are used to extend the existing Lyndon substrings where possible, or became Lyndon substrings of length one otherwise;*
(c) consolidate the new Lyndon substrings; *processed from the right, if several Lyndon substrings are adjacent and can be joined to a longer Lyndon substring, they are joined.*

The diagram in Figure 5 and the description that follows it illustrate the method for a string 011023122. The input string is visualized as a curve, and the height at each point is the value of the letter at that position.

In Figure 5, we illustrate the process:

(1) We start with the string 011023122 and a full tank of water.
(2) We drain one level; only 3 is revealed; there is nothing to extend, nothing to consolidate.
(3) We drain one more level, and three 2's are revealed; the first 2 extends 3 to 23, and the remaining two 2's form Lyndon substrings 2 of length one; there is nothing to consolidate.
(4) We drain one more level, and three 1's are revealed; the first two 1's form Lyndon substrings 1 of length one; the third 1 extends 22 to 122; there is nothing to consolidate.

(5)   We drain one more level, and two 0's are revealed; the first 0 extends 11 to 011; the second 0 extends 23 to 023; in the consolidation phase, 023 is joined with 122 to form a Lyndon substring 023122, and then, 011 is joined with 023122 to form a Lyndon substring 011023122.

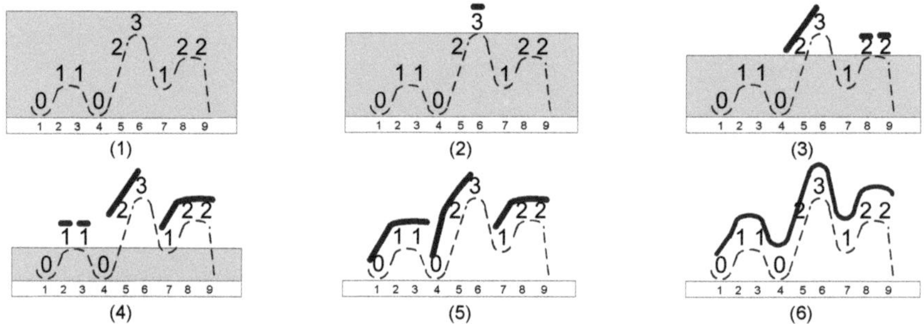

Figure 5. The water draining method for 011023122. Stages (1)–(6) explained in the text.

Therefore, during the process, the following right-maximal Lyndon substrings were identified: 3 at Position 6, 23 at Position 5, 2 at Positions 8 and 9, 1 at Positions 2 and 3, 122 at Position 7, 023 at Position 4, and finally, 011023122 at Position 1. Note that all positions are accounted for; we really have all right-maximal Lyndon substrings of the string 011023122.

In Figure 6, we present an illustrative example for the string 011023122, where the arrows represent the *prev* mapping shown only on the group used for the refinement. The groups used for the refinement are indicated by the bold font.

$$
\begin{array}{ccccccccc}
\mathbf{0} & \mathbf{1} & \mathbf{1} & \mathbf{0} & \mathbf{2} & \mathbf{3} & \mathbf{1} & \mathbf{2} & \mathbf{2} \\
1 & 2 & 3 & 4 & 5 & 6 & 7 & 8 & 9
\end{array}
$$

$G_0 = \{1,4\}\ G_1 = \{2,3,7\}\ G_2 = \{5,8,9\}\ G_3 = \{6\}$

$G_0 = \{1,4\}\ G_1 = \{2,3,7\}\ G_2 = \{8,9\}\ G_{23} = \{5\}\ G_3 = \{6\}$

$G_0 = \{1\}\ G_{023} = \{4\}\ G_1 = \{2,3,7\}\ G_2 = \{8,9\}\ G_{23} = \{5\}\ G_3 = \{6\}$

$G_0 = \{1\}\ G_{023} = \{4\}\ G_1 = \{2,3\}\ G_{122} = \{7\}\ G_2 = \{8,9\}\ G_{23} = \{5\}\ G_3 = \{6\}$

$G_0 = \{1\}\ G_{023122} = \{4\}\ G_1 = \{2,3\}\ G_{122} = \{7\}\ G_2 = \{8,9\}\ G_{23} = \{5\}\ G_3 = \{6\}$

$G_{011} = \{1\}\ G_{023122} = \{4\}\ G_1 = \{2,3\}\ G_{122} = \{7\}\ G_2 = \{8,9\}\ G_{23} = \{5\}\ G_3 = \{6\}$

$G_{011023122} = \{1\}\ G_{023122} = \{4\}\ G_1 = \{2,3\}\ G_{122} = \{7\}\ G_2 = \{8,9\}\ G_{23} = \{5\}\ G_3 = \{6\}$

Figure 6. Group refinement for 011023122.

4.4. The Complexity of BSLA

The computation of the initial configuration can be done in linear time. To compute the initial value of *prev* in linear time, a stack-based approach similar to the *NSV* algorithm is used. Since all groups are non-empty, there can never be more groups than n. Theorem 3 is at the heart of the algorithm. The refinement by the last completed group is linear in the size of the group, including the update of *prev*. Therefore, the overall worst-case complexity of *BSLA* is linear in the length of the input string.

5. Data and Measurements

Initially, computations were performed on the Department of Computing and Software's *moore* server; memory: 32 GB (DDR4 @ 2400 MHz), CPU: 8 × Intel Xeon E5-2687W v4 @ 3.00 GHz, OS: Linux Version 2.6.18-419.el5 (gcc Version 4.1.2 and Red Hat Version 4.1.2-55). To verify correctness, new randomized data were produced and computed independently on the University of Toronto Mississauga's *octolab* cluster; memory: 8 × 32 GB (DDR4 @ 3200 MHz), CPU: 8 × AMD Ryzen Threadripper 1920X (12-Core) @ 4.00 GHz, OS: Ubuntu 16.04.6 LTS (gcc Version 5.4.0). The results of both were extremely similar, and those reported herein are those generated using the *moore* server. All the programs were compiled without any additional level of optimization (i.e., neither -O1, nor -O2, nor -O3 flags were specified for the compilation). The CPU time was measured in clock ticks with 1,000,000 clock ticks per second. Since the execution time was negligible for short strings, the processing of the same string was repeated several times (the repeat factor varied from 10^6, for strings of length 10, to one, for strings of length 5×10^6), resulting in a higher precision. Thus, for graphing, the logarithmic scale was used for both, x-axis representing the length of the strings and y-axis representing the time. We used four categories of randomly generated datasets:

(1) `bin`
    random strings over an integer alphabet with exactly two distinct letters (kind of binary strings).
(2) `dna`
    random strings over an integer alphabet with exactly four distinct letters (kind of random DNA strings).
(3) `eng`
    random strings over an integer alphabet with exactly 26 distinct letters (kind of random English).
(4) `int`
    random strings over an integer alphabet (i.e., over the alphabet $\{0, ..., n-1\}$).

Each dataset contains 100 randomly generated strings of the same length. For each category, there were datasets for length s 10, 50, 10^2, 5×10^2, ..., 10^5, 5×10^5, 10^6, and 5×10^6. The minimum, average, and maximum times for each dataset were computed. Since the variance for each dataset was minimal, the results for minimum times and the results for maximum times completely mimicked the results for the average times, so we only present the averages here.

Tables 1–4 and the graphs in Figures 7–10 from Section 7 clearly indicate that the performance of the three algorithms is linear and virtually indistinguishable. We expected *IDLA* and *TRLA* to exhibit linear behaviour on random strings as such strings tend to have many, but short right-maximal Lyndon substrings. However, we did not expect the results to be so close.

Despite the fact that *IDLA* performed in linear time on the random strings, it is relatively easy to force it into its worst quadratic performance. The dataset `extreme_idla` contains individual strings $0123...n-1$ of the required lengths. Table 5 and the graph in Figure 11 from Section 7 show this clearly.

In Section 3.5, we describe how the three datasets, `extreme_trla`, `extreme_trla1`, and `extreme_trla2`, are generated and why. The results of experimenting with these datasets do not suggest that the worst-case complexity for *TRLA* is $\mathcal{O}(n \log(n))$. Yet again, the performances of the three algorithms are linear and virtually indistinguishable; see Tables 6–8 and the graphs in Figures 12–14 in Section 7.

6. Conclusions and Future Work

We present two novel algorithms for computing right-maximal Lyndon substrings. The first one, *TRLA*, has a simple implementation with a complicated theory behind it. Its average time complexity is linear in the length of the input string, and its worst-case complexity is no worse than $\mathcal{O}(n \log(n))$.

The τ-reduction used in the algorithm is an interesting reduction preserving right-maximal Lyndon substrings, a fact used significantly in the design of the algorithm. Interestingly, it seem to slightly outperform *BSLA*, at least on the datasets used for our experimentations. *BSLA*, the second algorithm, is linear and elementary in the sense that it does not require a pre-processed global data structure. Being linear and elementary, *BSLA* is more interesting, and it is possible that its performance could be more streamlined. However, both the theory and implementation of *BSLA* are rather complex.

On random strings, none of the two algorithms were significantly better than the simple *IDLA*, whose implementation is just a few lines. However, its quadratic worst-case complexity is an obstacle, as our experiments indicated.

Additional effort needs to go into proving *TRLA*'s worst-case complexity. The experiments performed did not indicate that it is not linear even in the worst case. Both algorithms need to be compared to some efficient implementation of *SSLA* and *BWLA*.

7. Results

This section contains the measurements of the average times for the datasets discussed in the previous section. For better understanding of the data, we present them in Tables 1–8 and Figures 7–14. All the graphs include the curve $x = y$ for reference.

Table 1. Average times for dataset `bin` (10^6 *clock ticks per second*).

String Length	Time in Ticks IDLA	Time in Ticks TRLA	Time in Ticks BSLA
10	3.651×10^{-1}	1.582	1.054
50	4.082	1.050×10	6.372
100	1.101×10	2.140×10	1.277×10
500	8.655×10	1.127×10^2	6.786×10
1000	1.975×10^2	2.335×10^2	1.484×10^2
5000	1.278×10^3	1.218×10^3	8.595×10^2
10,000	2.765×10^3	2.423×10^3	1.820×10^3
50,000	1.665×10^4	1.272×10^4	1.018×10^4
100,000	3.606×10^4	2.523×10^4	2.113×10^4
500,000	2.071×10^5	1.338×10^5	1.493×10^5
1,000,000	4.387×10^5	2.717×10^5	4.080×10^5
5,000,000	2.483×10^6	1.561×10^6	3.098×10^6

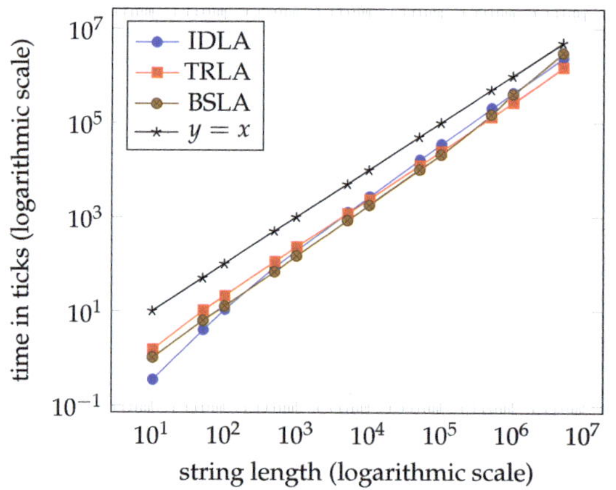

Figure 7. Average times for dataset `bin` (10^6 *clock ticks per second*).

Table 2. Average times for dataset dna (10^6 *clock ticks per second*).

String Length	Time in Ticks IDLA	Time in Ticks TRLA	Time in Ticks BSLA
10	3.699×10^{-1}	1.579	1.080
50	3.509	1.037×10	6.627
100	8.898	2.109×10	1.403×10
500	6.403×10	1.123×10^2	7.228×10
1000	1.431×10^2	2.332×10^2	1.544×10^2
5000	8.749×10^2	1.207×10^3	9.039×10^2
10,000	1.912×10^3	2.460×10^3	1.935×10^3
50,000	1.134×10^4	1.280×10^4	1.110×10^4
100,000	2.431×10^4	2.588×10^4	2.316×10^4
500,000	1.383×10^5	1.390×10^5	1.781×10^5
1,000,000	2.916×10^5	2.865×10^5	4.994×10^5
5,000,000	1.643×10^6	1.968×10^6	3.752×10^6

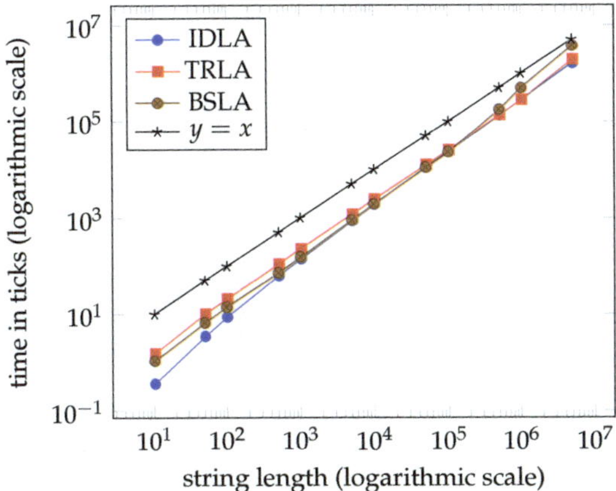

Figure 8. Average times for dataset dna (10^6 *clock ticks per second*).

Table 3. Average times for dataset eng (10^6 *clock ticks per second*).

String Length	Time in Ticks IDLA	Time in Ticks TRLA	Time in Ticks BSLA
10	3.526×10^{-1}	1.584	9.865×10^{-1}
50	3.162	1.006×10	5.960
100	7.315	2.057×10	1.317×10
500	4.996×10	1.117×10^2	7.245×10
1000	1.112×10^2	2.354×10^2	1.542×10^2
5000	6.722×10^2	1.210×10^3	9.087×10^2
10,000	1.452×10^3	2.427×10^3	2.042×10^3
50,000	8.505×10^3	1.306×10^4	1.301×10^4
100,000	1.802×10^4	2.688×10^4	2.768×10^4
500,000	1.025×10^5	1.428×10^5	2.381×10^5
1,000,000	2.171×10^5	3.253×10^5	7.236×10^5
5,000,000	1.206×10^6	2.599×10^6	6.092×10^6

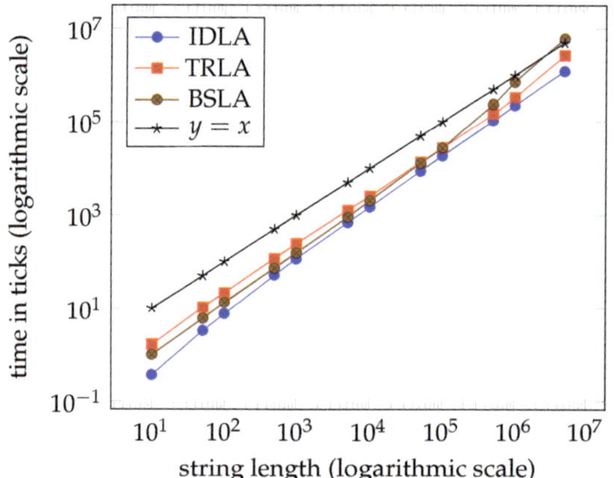

Figure 9. Average times for dataset `eng` (10^6 *clock ticks per second*).

Table 4. Average times for dataset `int` (10^6 *clock ticks per second*).

String Length	Time in Ticks IDLA	Time in Ticks TRLA	Time in Ticks BSLA
10	3.547×10^{-1}	1.645	9.794×10^{-1}
50	3.032	9.992	5.609
100	7.279	2.032×10	1.153×10
500	4.845×10	1.136×10^2	6.184×10
1000	1.057×10^2	2.376×10^2	1.294×10^2
5000	6.428×10^2	1.218×10^3	7.753×10^2
10,000	1.388×10^3	2.544×10^3	1.796×10^3
50,000	8.055×10^3	1.448×10^4	1.088×10^4
100,000	1.710×10^4	2.943×10^4	2.379×10^4
500,000	9.829×10^4	1.825×10^5	2.740×10^5
1,000,000	2.071×10^5	4.827×10^5	7.989×10^5
5,000,000	1.162×10^6	5.143×10^6	6.635×10^6

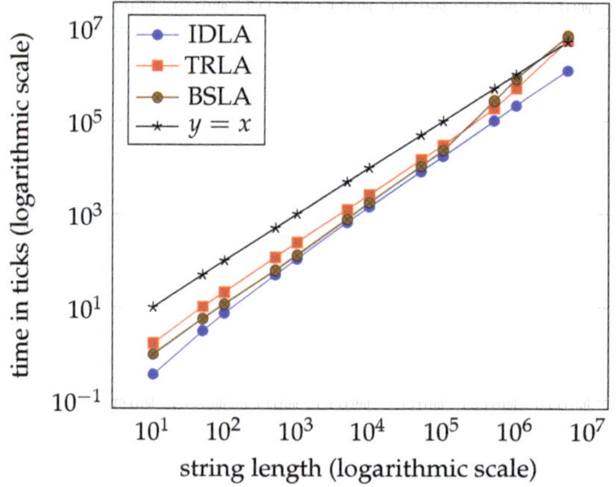

Figure 10. Average times for dataset `int` (10^6 *clock ticks per second*).

Table 5. Average times for dataset `extreme_idla` (10^6 *clock ticks per second*).

String Length	Time in Ticks IDLA	Time in Ticks TRLA	Time in Ticks BSLA
10	7.900×10^{-1}	1.440	7.000×10^{-1}
50	1.830×10	8.200	3.600
100	7.190×10	1.590×10	6.800
500	1.778×10^3	7.300×10	3.550×10
1000	7.105×10^3	1.430×10^2	6.900×10
5000	1.776×10^5	7.100×10^2	3.400×10^2
10,000	7.111×10^5	1.550×10^3	6.800×10^2
50,000	1.784×10^7	8.050×10^3	3.400×10^3
100,000	7.130×10^7	1.600×10^4	6.800×10^3
500,000	1.783×10^9	8.200×10^4	3.700×10^4
1,000,000	7.137×10^9	1.660×10^5	7.800×10^4
5,000,000	1.813×10^{11}	8.800×10^5	4.950×10^5

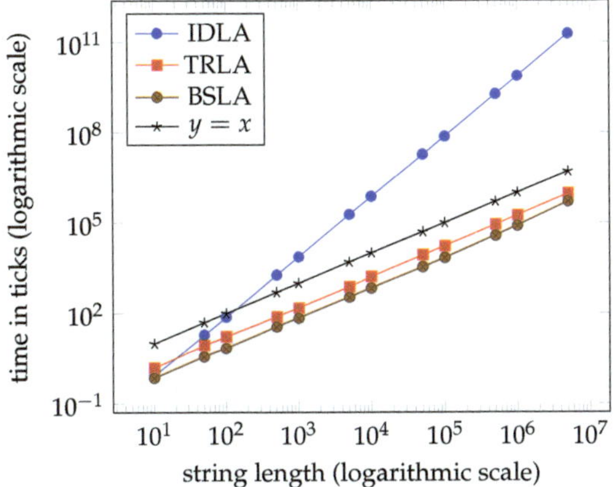

Figure 11. Average times for dataset `extreme_idla` (10^6 *clock ticks per second*).

Table 6. Average times for dataset `extreme_trla` (10^6 *clock ticks per second*).

String Length	Time in Ticks IDLA	Time in Ticks TRLA	Time in Ticks BSLA
10	4.588×10^{-1}	1.628	1.126
50	4.987	1.039×10	7.112
100	1.275×10	2.179×10	1.439×10
500	9.033×10	1.101×10^2	6.914×10
1000	2.060×10^2	2.222×10^2	1.392×10^2
5000	1.319×10^3	1.171×10^3	7.699×10^2
10,000	2.896×10^3	2.394×10^3	1.652×10^3
50,000	2.209×10^4	1.263×10^4	8.992×10^3
100,000	3.965×10^4	2.567×10^4	1.862×10^4
500,000	2.233×10^5	1.349×10^5	1.091×10^5
1,000,000	4.734×10^5	2.759×10^5	3.104×10^5
5,000,000	2.632×10^6	1.458×10^6	2.298×10^6

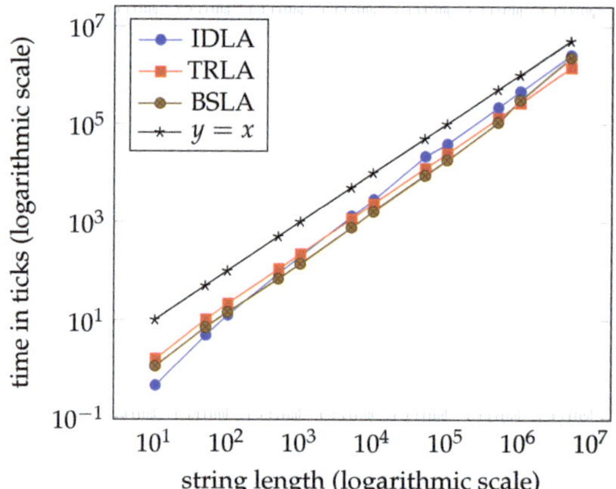

Figure 12. Average times for dataset `extreme_trla` (10^6 *clock ticks per second*).

Table 7. Average times for dataset `extreme_trla1` (10^6 *clock ticks per second*).

String Length	Time in Ticks IDLA	Time in Ticks TRLA	Time in Ticks BSLA
10	5.040×10^{-1}	1.600	1.117
50	5.910	1.042×10	7.290
100	1.460×10	2.145×10	1.446×10
500	1.146×10^2	1.126×10^2	6.979×10
1000	2.662×10^2	2.284×10^2	1.379×10^2
5000	1.694×10^3	1.205×10^3	7.853×10^2
10,000	3.734×10^3	2.477×10^3	1.739×10^3
50,000	2.276×10^4	1.310×10^4	9.683×10^3
100,000	4.901×10^4	2.796×10^4	2.009×10^4
500,000	2.928×10^5	1.465×10^5	1.238×10^5
1,000,000	6.199×10^5	3.000×10^5	3.622×10^5
5,000,000	3.432×10^6	1.642×10^6	2.778×10^6

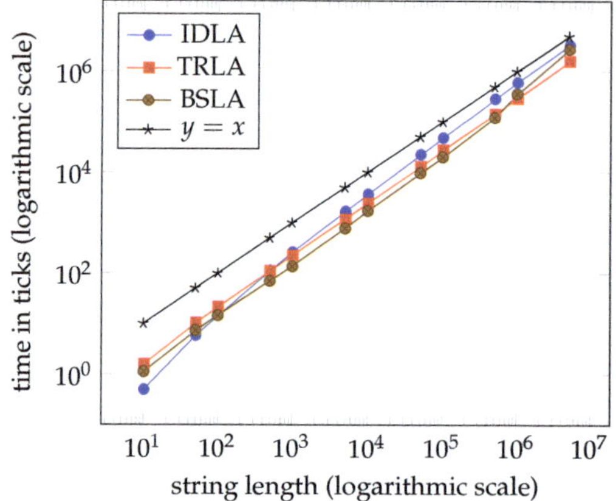

Figure 13. Average times for dataset `extreme_trla1` (10^6 *clock ticks per second*).

Table 8. Average times for dataset `extreme_trla2` (10^6 *clock ticks per second*).

String Length	Time in Ticks IDLA	Time in Ticks TRLA	Time in Ticks BSLA
10	5.041×10^{-1}	1.683	1.121
50	6.160	1.020×10	7.257
100	1.526×10	2.090×10	1.441×10
500	1.367×10^2	1.074×10^2	7.117×10
1000	3.202×10^2	2.135×10^2	1.390×10^2
5000	2.024×10^3	1.145×10^3	7.966×10^2
10,000	4.500×10^3	2.257×10^3	1.762×10^3
50,000	2.728×10^4	1.172×10^4	1.012×10^4
100,000	5.941×10^4	2.362×10^4	2.115×10^4
500,000	3.639×10^5	1.262×10^5	1.351×10^5
1,000,000	7.719×10^5	2.571×10^5	3.915×10^5
5,000,000	4.263×10^6	1.323×10^6	3.118×10^6

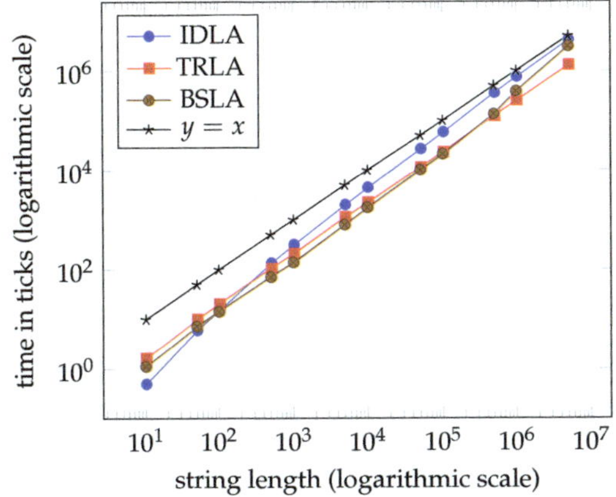

Figure 14. Average times for dataset `extreme_trla2` (10^6 *clock ticks per second*).

Author Contributions: Conceptualization, F.F. and M.L.; methodology, F.F. and M.L.; software, F.F. and M.L.; validation, F.F. and M.L.; formal analysis, F.F. and M.L.; investigation, F.F. and M.L.; resources, F.F. and M.L.; data curation, F.F. and M.L.; writing, original draft preparation, F.F. and M.L.; writing, review and editing, F.F. and M.L.; visualization, F.F. and M.L.; supervision, F.F. and M.L.; project administration, F.F. and M.L.; funding acquisition, F.F. Both authors read and agreed to the published version of the manuscript.

Funding: This research was funded by the National Sciences and Research Council of Canada (NSERC) Grant RGPIN/5504-2018.

Conflicts of Interest: The authors declare no conflict of interest.

Abbreviations

The following abbreviations are used in this manuscript:

BSLA	Baier's Sort Lyndon Array
IDLA	Iterative Duval Lyndon Array
TRLA	Tau Reduction Lyndon Array

References

1. Lyndon, R.C. On Burnside's Problem. II. *Trans. Am. Math. Soc.* **1955**, *78*, 329–332.
2. Marcus, S.; Sokol, D. 2D Lyndon words and applications. *Algorithmica* **2017**, *77*, 116–133. [CrossRef]
3. Berstel, J.; Perrin, D. The origins of combinatorics on words. *Eur. J. Comb.* **2007**, *28*, 996–1022. [CrossRef]

4. Chen, K.; Fox, R.; Lyndon, R. Free differential calculus IV. The quotient groups of the lower central series. *Ann. Math. 2nd Ser.* **1958**, *68*, 81–95. [CrossRef]

5. Golomb, S. Irreducible polynomials, synchronizing codes, primitive necklaces and cyclotomic algebra. *Comb. Math. Appl.* **1967**, *4*, 358–370.

6. Flajolet, P.; Gourdon, X.; Panario, D. The complete analysis of a polynomial factorization algorithm over finite fields. *J. Algorithms* **2001**, *40*, 37–81. [CrossRef]

7. Panario, D.; Richmond, B. Smallest components in decomposable structures:exp-log class. *Algorithmica* **2001**, *29*, 205–226. [CrossRef]

8. Duval, J.P. Factorizing words over an ordered alphabet. *J. Algorithms* **1983**, *4*, 363–381. [CrossRef]

9. Berstel, J.; Pocchiola, M. Average cost of Duval's algorithm for generating Lyndon words. *Theor. Comput. Sci.* **1994**, *132*, 415–425. [CrossRef]

10. Fredricksen, H.; Maiorana, J. Necklaces of beads in *k* colors and *k*-ary de Bruijn sequences. *Discret. Math.* **1983**, *23*, 207–210. [CrossRef]

11. Bannai, H.; Tomohiro, I.; Inenaga, S.; Nakashima, Y.; Takeda, M.; Tsuruta, K. The "Runs" Theorem. *SIAM J. Comput.* **2017**, *46*, 1501–1514. [CrossRef]

12. Franek, F.; Paracha, A.; Smyth, W. The linear equivalence of the suffix array and the partially sorted Lyndon array. In Proceedings of the Prague Stringology Conference, Prague, Czech Republic, 28–30 August 2017; pp. 77–84.

13. Baier, U. Linear-Time Suffix Sorting—A New Approach for Suffix Array Construction. Master's Thesis, University of Ulm, Ulm, Germany, 2015.

14. Baier, U. Linear-Time Suffix Sorting—A New Approach for Suffix Array Construction. In Proceedings of the 27th Annual Symposium on Combinatorial Pattern Matching (CPM 2016), Tel Aviv, Israel, 27–29 June 2016; Grossi, R., Lewenstein, M., Eds.; Schloss Dagstuhl–Leibniz-Zentrum fuer Informatik: Dagstuhl, Germany, 2016; Volume 54, pp. 1–12.

15. Chen, G.; Puglisi, S.; Smyth, W. Lempel-Ziv factorization using less time & space. *Math. Comput. Sci.* **2013**, *1*, 605–623.

16. Crochemore, M.; Ilie, L.; Smyth, W. A simple algorithm for computing the Lempel-Ziv factorization. In Proceedings of the 18th Data Compression Conference, Snowbird, UT, USA, 25–27 March 2008; pp. 482–488.

17. Kosolobov, D. Lempel-Ziv factorization may be harder than computing all runs. In Proceedings of the 32 International Symposium on Theoretical Aspects of Computer Science—STACS 2015, Garching, Germany, 4–7 March 2015; pp. 582–593.

18. Digelmann, C. (Frankfurt, Germany). Personal communication, 2016.

19. Franek, F.; Sohidull Islam, A.; Sohel Rahman, M.; Smyth, W. Algorithms to compute the Lyndon array. In Proceedings of the Prague Stringology Conference 2016, Prague, Czech Republic, 29–31 August 2016; pp. 172–184.

20. Hohlweg, C.; Reutenauer, C. Lyndon words, permutations and trees. *Theor. Comput. Sci.* **2003**, *307*, 173–178. [CrossRef]

21. Nong, G.; Zhang, S.; Chan, W.H. Linear suffix array construction by almost pure induced-sorting. In Proceedings of the 2009 Data Compression Conference, Snowbird, UT, USA, 16–18 March 2009; pp. 193–202.

22. Louza, F.; Smyth, W.; Manzini, G.; Telles, G. Lyndon array construction during Burrows–Wheeler inversion. *J. Discret. Algorithms* **2018**, *50*, 2–9. [CrossRef]

23. Franek, F.; Liut, M.; Smyth, W. On Baier's sort of maximal Lyndon substrings. In Proceedings of the Prague Stringology Conference 2018, Prague, Czech Republic, 27–28 August 2018; pp. 63–78.

24. C++ Code for IDLA, TRLA and BSLA Algorithms. Available online: https://github.com/MichaelLiut/Computing-LyndonArray (accessed on 3 November 2020).

25. Farach, M. Optimal suffix tree construction with large alphabets. In Proceedings of the 38th IEEE Symp. Foundations of Computer Science, Miami Beach, FL, USA, 20–22 October 1997; pp. 137–143.

26. Nong, G. Practical linear-time $O(1)$-workspace suffix sorting for constant alphabets. *ACM Trans. Inf. Syst.* **2013**, *31*, 1–15. [CrossRef]

27. Cooley, J.; Tukey, J. An algorithm for the machine calculation of complex Fourier series. *Math. Comput.* **1965**, *19*, 297–301. [CrossRef]

28. Franek, F.; Liut, M. Computing Maximal Lyndon Substrings of a String, AdvOL Report 2019/2, McMaster University. Available online: http://optlab.mcmaster.ca//component/option,com_docman/task,cat_view/gid,77/Itemid,92 (accessed on 1 March 2019).

29. Franek, F.; Liut, M. Algorithms to compute the Lyndon array revisited. In Proceedings of the Prague Stringology Conference 2019, Prague, Czech Republic, 26–28 August 2019; pp. 16–28.

30. Liut, M. Computing Lyndon Arrays. Ph.D. Thesis, McMaster University, Hamilton, ON, Canada, 2019.

31. Lothaire, M. *Combinatorics on Words*; Cambridge University Press: Cambridge, UK, 2003.

32. Lothaire, M. *Applied Combinatorics on Words*; Cambridge University Press: Cambridge, UK, 2005.

33. Smyth, B. *Computing Patterns in Strings*; Pearson Addison-Wesley: Boston, MA, USA, 2003.

34. Louza, F.; Gog, S.; Telles, G. *Construction of Fundamental Data Structures for Strings*; Springer: Cham, Switzerland, 2020.

35. Burkhardt, S.; Kärkkäinen, J. Fast Lightweight Suffix Array Construction and Checking. In Proceedings of the 14th Annual Conference on Combinatorial Pattern Matching, Michoacan, Mexico, 25–27 June 2003; Springer: Berlin, Heidelberg, 2003; pp. 55–69.

36. Paracha, A. Lyndon Factors and Periodicities in Strings. Ph.D. Thesis, McMaster University, Hamilton, ON, Canada, 2017.

37. Kärkkäinen, J.; Sanders, P. Simple linear work suffix array construction. In Proceedings of the 30th International Conference on Automata, Languages and Programming, Eindhoven, The Netherlands, 30 June–4 July 2003; Springer: Berlin/Heidelberg, Germany, 2003; pp. 943–955.

Publisher's Note: MDPI stays neutral with regard to jurisdictional claims in published maps and institutional affiliations.

 *algorithms*

Article

Re-Pair in Small Space [†]

Dominik Köppl [1,*] [iD], **Tomohiro I** [2] [iD], **Isamu Furuya** [3] [iD], **Yoshimasa Takabatake** [2] [iD], **Kensuke Sakai** [2] and **Keisuke Goto** [4]

[1] M&D Data Science Center, Tokyo Medical and Dental University, Tokyo 113-8510, Japan
[2] Kyushu Institute of Technology, Fukuoka 820-8502, Japan; tomohiro@ai.kyutech.ac.jp (T.I.);
     takabatake@ai.kyutech.ac.jp (Y.T.); k_sakai@donald.ai.kyutech.ac.jp (K.S.)
[3] Graduate School of IST, Hokkaido University, Hokkaido 060-0814, Japan; furuya@ist.hokudai.ac.jp
[4] Fujitsu Laboratories Ltd., Kawasaki 211-8588, Japan; goto.keisuke@fujitsu.com
[*] Correspondence: koeppl.dsc@tmd.ac.jp; Tel.: +81-3-5280-8626
[†] This paper is an extended version of our paper published in the Prague Stringology Conference 2020: Prague, Czech Republic, 31 August–2 September 2020 and at the Data Compression Conference 2020: Virtual Conference, 24–27 March 2020.

Abstract: Re-Pair is a grammar compression scheme with favorably good compression rates. The computation of Re-Pair comes with the cost of maintaining large frequency tables, which makes it hard to compute Re-Pair on large-scale data sets. As a solution for this problem, we present, given a text of length n whose characters are drawn from an integer alphabet with size $\sigma = n^{\mathcal{O}(1)}$, an $\mathcal{O}(\min(n^2, n^2 \lg \log_\tau n \lg \lg \lg n / \log_\tau n))$ time algorithm computing Re-Pair with $\max((n/c) \lg n, n \lceil \lg \tau \rceil) + \mathcal{O}(\lg n)$ bits of working space including the text space, where $c \geq 1$ is a fixed user-defined constant and τ is the sum of σ and the number of non-terminals. We give variants of our solution working in parallel or in the external memory model. Unfortunately, the algorithm seems not practical since a preliminary version already needs roughly one hour for computing Re-Pair on one megabyte of text.

Keywords: grammar compression; Re-Pair; computation in small space; broadword techniques

Citation: Köppl, D.; I, T.; Furuya, I.; Takabatake, Y.; Sakai, K.; Goto, K. Re-Pair in Small Space. *Algorithms* **2021**, *14*, 5. https://dx.doi.org/10.3390/a14010005

Received: 29 November 2020
Accepted: 18 December 2020
Published: 25 December 2020

Publisher's Note: MDPI stays neutral with regard to jurisdictional claims in published maps and institutional affiliations.

1. Introduction

Re-Pair [1] is a grammar deriving a single string. It is computed by replacing the most frequent bigram in this string with a new non-terminal, recursing until no bigram occurs more than once. Despite this simple-looking description, both the merits and the computational complexity of Re-Pair are intriguing. As a matter of fact, Re-Pair is currently one of the most well-understood grammar schemes.

Besides the seminal work of Larsson and Moffat [1], there are a couple of articles devoted to the compression aspects of Re-Pair: Given a text T of length n whose characters are drawn from an integer alphabet of size $\sigma := n^{\mathcal{O}(1)}$, the output of Re-Pair applied to T is at most $2nH_k(T) + o(n \lg \sigma)$ bits with $k = o(\log_\sigma n)$ when represented naively as a list of character pairs [2], where H_k denotes the empirical entropy of the k-th order. Using the encoding of Kieffer and Yang [3], Ochoa and Navarro [4] could improve the output size to at most $nH_k(T) + o(n \lg \sigma)$ bits. Other encodings were recently studied by Ganczorz [5]. Since Re-Pair is a so-called irreducible grammar, its grammar size, i.e., the sum of the symbols on the right-hand side of all rules, is upper bounded by $\mathcal{O}(n/\log_\sigma n)$ ([3], Lemma 2), which matches the information-theoretic lower bound on the size of a grammar for a string of length n. Comparing this size with the size of the smallest grammar, its approximation ratio has $\mathcal{O}((n/\lg n)^{2/3})$ as an upper bound [6] and $\Omega(\lg n/\lg \lg n)$ as a lower bound [7]. On the practical side, Yoshida and Kida [8] presented an efficient fixed-length code for compressing the Re-Pair grammar.

Although conceived of as a grammar for compressing texts, Re-Pair has been successfully applied for compressing trees [9], matrices [10], or images [11]. For different settings or for better compression rates, there is a great interest in modifications to Re-Pair. Charikar et al. [6] (Section G) gave an easy variation to improve the size of the grammar. Another variant, proposed by Claude and Navarro [12], runs in a user-defined working space ($> n \lg n$ bits) and shares with our proposed solution the idea of a table that (a) is stored with the text in the working space and (b) grows in rounds. The variant of González et al. [13] is specialized to compressing a delta-encoded array of integers (i.e., by the differences of subsequent entries). Sekine et al. [14] provided an adaptive variant whose algorithm divides the input into blocks and processes each block based on the rules obtained from the grammars of its preceding blocks. Subsequently, Masaki and Kida [15] gave an online algorithm producing a grammar mimicking Re-Pair. Ganczorz and Jez [16] modified the Re-Pair grammar by disfavoring the replacement of bigrams that cross Lempel–Ziv-77 (LZ77) [17] factorization borders, which allowed the authors to achieve practically smaller grammar sizes. Recently, Furuya et al. [18] presented a variant, called MR-Re-Pair, in which a most frequent maximal repeat is replaced instead of a most frequent bigram.

1.1. Related Work

In this article, we focus on the problem of computing the grammar with an algorithm working in text space, forming a bridge between the domain of in-place string algorithms, low-memory compression algorithms, and the domain of Re-Pair computing algorithms. We briefly review some prominent achievements in both domains:

In-place string algorithms: For the LZ77 factorization, Kärkkäinen et al. [19] presented an algorithm computing this factorization with $\mathcal{O}(n/d)$ words on top of the input space in $\mathcal{O}(dn)$ time for a variable $d \geq 1$, achieving $\mathcal{O}(1)$ words with $\mathcal{O}(n^2)$ time. For the suffix sorting problem, Goto [20] gave an algorithm to compute the suffix array [21] with $\mathcal{O}(\lg n)$ bits on top of the output in $\mathcal{O}(n)$ time if each character of the alphabet is present in the text. This condition was improved to alphabet sizes of at most n by Li et al. [22]. Finally, Crochemore et al. [23] showed how to transform a text into its Burrows–Wheeler transform by using $\mathcal{O}(\lg n)$ of additional bits. Due to da Louza et al. [24], this algorithm was extended to compute simultaneously the longest common prefix (LCP) array [21] with $\mathcal{O}(\lg n)$ bits of additional working space.

Low-memory compression algorithms: Simple compression algorithms like run-length compression can be computed in-place and online on the text in linear time. However, a similar result for LZ77 is unknown: A trivial algorithm working with constant number of words (omitting the input text) computes an LZ77 factor starting at $T[i..]$ by linearly scanning $T[1..i-1]$ for the longest previous occurrence $T[j..j+\ell-1] = T[i..i+\ell-1]$ for $j < i$, thus taking quadratic time. A trade-off was proposed by Kärkkäinen et al. [19], who needed $\mathcal{O}(n \lg n/d)$ bits of working space and $\mathcal{O}(nd \lg \lg_n \sigma)$ time for a selectable parameter $d \geq 1$. For the particular case of $d = \epsilon^{-1} \lg n$ for an arbitrary constant $\epsilon > 0$, Kosolobov [25] could improve the running time to $\mathcal{O}(n(\lg \sigma + \lg((\lg n)/\epsilon))/\epsilon)$ for the same space of $\mathcal{O}(\epsilon n)$ bits. Unfortunately, we are unaware of memory-efficient algorithms computing other grammars such as longest-first substitution (LFS) [26], where a modifiable suffix tree is used for computation.

Re-Pair computation: Re-Pair is a grammar proposed by Larsson and Moffat [1], who presented an algorithm computing it in expected linear time with $5n + 4\sigma^2 + 4\sigma' + \sqrt{n}$ words of working space, where σ' is the number of non-terminals (produced by Re-Pair). González et al. [13] (Section 4.1) gave another linear time algorithm using $12n + \mathcal{O}(p)$ bytes of working space, where p is the maximum number of distinct bigrams considered at any time. The large space requirements got significantly improved by Bille et al. [27], who presented a randomized linear time algorithm taking $(1 + \epsilon)n + \sqrt{n}$ words on top of the rewritable text space for a constant ϵ with $0 < \epsilon \leq 1$. Subsequently, they improved their algorithm in [28] to include the text space within the $(1 + \epsilon)n + \sqrt{n}$ words of the working space. However, they assumed that the alphabet size σ was constant and $\lceil \lg \sigma \rceil \leq w/2$, where w is the machine word size. They also provided a solution for $\epsilon = 0$ running in expected linear time. Recently, Sakai et al. [29] showed how to convert an arbitrary grammar (representing a text) into the Re-Pair grammar in compressed space, i.e., without decompressing the text. Combined with a grammar compression that can process the text in compressed space in a streaming fashion, this result leads to the first Re-Pair computation in compressed space.

In a broader picture, Carrascosa et al. [30] provided a generalization called iterative repeat replacement (IRR) , which iteratively selects a substring for replacement via a scoring function. Here, Re-Pair and its variant MR-Re-Pair are specializations of the provided grammar IRR-MO (IRR with maximal number of occurrences)selecting one of the most frequent substrings that have a reoccurring non-overlapping occurrence. (As with bigrams, we only count the number of non-overlapping occurrences.)

1.2. Our Contribution

In this article, we propose an algorithm that computes the Re-Pair grammar in $\mathcal{O}(\min(n^2, n^2 \lg \log_\tau n \lg \lg \lg n / \log_\tau n))$ time (cf. Theorems 1 and 2) with $\max((n/c) \lg n, n\lceil \lg \tau \rceil) + \mathcal{O}(\lg n)$ bits of working space including the text space, where $c \geq 1$ is a fixed user-defined constant and τ is the sum of the alphabet size σ and the number of non-terminals σ'.

We can also compute the byte pair encoding [31], which is Re-Pair with the additional restriction that the algorithm terminates before $\lceil \lg \tau \rceil = \lceil \lg \sigma \rceil$ no longer holds. Hence, we can replace τ with σ in the above space and time bounds.

Given that the characters of the text are drawn from a large integer alphabet with size $\sigma = \Omega(n)$ the algorithm works in-place. (We consider the alphabet as not effective, i.e., a character does not have to appear in the text, as this is a common setting in Unicode texts such as Japanese text. For instance, $n^2 = \Omega(n) \cap n^{\mathcal{O}(1)} \neq \varnothing$ could be such an alphabet size.) In this setting, we obtain the first non-trivial in-place algorithm, as a trivial approach on a text T of length n would compute the most frequent bigram in $\Theta(n^2)$ time by computing the frequency of each bigram $T[i]T[i+1]$ for every integer i with $1 \leq i \leq n-1$, keeping only the most frequent bigram in memory. This sums up to $\mathcal{O}(n^3)$ total time and can be $\Theta(n^3)$ for some texts since there can be $\Theta(n)$ different bigrams considered for replacement by Re-Pair.

To achieve our goal of $\mathcal{O}(n^2)$ total time, we first provide a trade-off algorithm (cf. Lemma 2) finding the d most frequent bigrams in $\mathcal{O}(n^2 \lg d/d)$ time for a trade-off parameter d. We subsequently run this algorithm for increasing values of d and show that we need to run it $\mathcal{O}(\lg n)$ times, which gives us $\mathcal{O}(n^2)$ time if d is increasing sufficiently fast. Our major tools are appropriate text partitioning, elementary scans, and sorting steps, which we visualize in Section 2.5 by an example and practically evaluate in Section 2.6. When $\tau = o(n)$, a different approach using word-packing and bit-parallel techniques becomes attractive, leading to an $\mathcal{O}(n^2 \lg \log_\tau n \lg \lg \lg n / \log_\tau n)$ time algorithm, which we explain in Section 3. Our algorithm can be parallelized (Section 5), used in external memory (Section 6), or adapted to compute the MR-Re-Pair grammar in small space (Section 4). Finally, in Section 7, we study several heuristics that make the algorithm faster on specific texts.

1.3. Preliminaries

We use the word RAM model with a word size of $\Omega(\lg n)$ for an integer $n \geq 1$. We work in the restore model [32], in which algorithms are allowed to overwrite the input, as long as they can restore the input to its original form.

Strings: Let T be a text of length n whose characters are drawn from an integer alphabet Σ of size $\sigma = n^{\mathcal{O}(1)}$. A bigram is an element of Σ^2. The *frequency* of a bigram B in T is the number of non-overlapping occurrences of B in T, which is at most $|T|/2$. For instance, the frequency of the bigram $\mathrm{aa} \in \Sigma^2$ in the text $T = \mathrm{a} \cdots \mathrm{a}$ consisting of n a's is $\lfloor n/2 \rfloor$.

Re-Pair: We reformulate the recursive description in the Introduction by dividing a Re-Pair construction algorithm into turns. Stipulating that T_i is the text after the i-th turn with $i \geq 1$ and $T_0 := T \in \Sigma_0^+$ with $\Sigma_0 := \Sigma$, Re-Pair replaces one of the most frequent bigrams (ties are broken arbitrarily) in T_{i-1} with a non-terminal in the i-th turn. Given this bigram is $\mathrm{bc} \in \Sigma_{i-1}^2$, Re-Pair replaces all occurrences of bc with a new non-terminal X_i in T_{i-1} and sets $\Sigma_i := \Sigma_{i-1} \cup \{X_i\}$ with $\sigma_i := |\Sigma_i|$ to produce $T_i \in \Sigma_i^+$. Since $|T_i| \leq |T_{i-1}| - 2$, Re-Pair terminates after $m < n/2$ turns such that $T_m \in \Sigma_m^+$ contains no bigram occurring more than once.

2. Sequential Algorithm

A major task for producing the Re-Pair grammar is to count the frequencies of the most frequent bigrams. Our work horse for this task is a frequency table. A *frequency table* in T_i of length f stores pairs of the form (bc, x), where bc is a bigram and x the frequency of bc in T_i. It uses $f\lceil \lg(\sigma_i^2 n_i/2)\rceil$ bits of space since an entry stores a bigram consisting of two characters from Σ_i and its respective frequency, which can be at most $n_i/2$. Throughout this paper, we use an elementary in-place sorting algorithm like heapsort:

Lemma 1 ([33]). *An array of length n can be sorted in-place in $\mathcal{O}(n \lg n)$ time.*

2.1. Trade-Off Computation

Using the frequency tables, we present a solution with a trade-off parameter:

Lemma 2. *Given an integer d with $d \geq 1$, we can compute the frequencies of the d most frequent bigrams in a text of length n whose characters are drawn from an alphabet of size σ in $\mathcal{O}(\max(n,d)n \lg d/d)$ time using $2d\lceil \lg(\sigma^2 n/2)\rceil + \mathcal{O}(\lg n)$ bits.*

Proof. Our idea is to partition the set of all bigrams appearing in T into $\lceil n/d \rceil$ subsets, compute the frequencies for each subset, and finally, merge these frequencies. In detail, we partition the text $T = S_1 \cdots S_{\lceil n/d \rceil}$ into $\lceil n/d \rceil$ substrings such that each substring has length d (the last one has a length of at most d). Subsequently, we extend S_j to the left (only if $j > 1$) such that S_j and S_{j+1} overlap by one text position, for $1 \leq j < \lceil n/d \rceil$. By doing so, we take the bigram on the border of two adjacent substrings S_j and S_{j+1} for each $j < \lceil n/d \rceil$ into account. Next, we create two frequency tables F and F', each of length d for storing the frequencies of d bigrams. These tables are at the beginning empty. In what follows, we fill F such that after processing S_i, F stores the most frequent d bigrams among those bigrams occurring in $S_1, \ldots, S_i$ while F' acts as a temporary space for storing candidate bigrams that can enter F.

With F and F', we process each of the n/d substrings S_j as follows: Let us fix an integer j with $1 \leq j \leq \lceil n/d \rceil$. We first put all bigrams of S_j into F' in lexicographic order. We can perform this within the space of F' in $\mathcal{O}(d \lg d)$ time since there are at most d different bigrams in S_j. We compute the frequencies of all these bigrams in the complete text T in $\mathcal{O}(n \lg d)$ time by scanning the text from left to right while locating a bigram in F' in $\mathcal{O}(\lg d)$ time with a binary search. Subsequently, we interpret F and F' as one large frequency table, sort it with respect to the frequencies while discarding duplicates such that F stores the d most frequent bigrams in $T[1..jd]$. This sorting step can be done in $\mathcal{O}(d \lg d)$ time. Finally, we clear F' and are done with S_j. After the final merge step, we obtain the d most frequent bigrams of T stored in F.

Since each of the $\mathcal{O}(n/d)$ merge steps takes $\mathcal{O}(d \lg d + n \lg d)$ time, we need: $\mathcal{O}(\max(d,n) \cdot (n \lg d)/d)$ time. For $d \geq n$, we can build a large frequency table and perform one scan to count the frequencies of all bigrams in T. This scan and the final sorting with respect to the counted frequencies can be done in $\mathcal{O}(n \lg n)$ time. $\square$

2.2. Algorithmic Ideas

With Lemma 2, we can compute T_m in $\mathcal{O}(mn^2 \lg d/d)$ time with additional $2d\lceil \lg(\sigma_m^2 n/2)\rceil$ bits of working space on top of the text for a parameter d with $1 \leq d \leq n$. (The variable τ

used in the abstract and in the introduction is interchangeable with σ_m, i.e., $\tau = \sigma_m$.) In the following, we show how this leads us to our first algorithm computing Re-Pair:

Theorem 1. *We can compute Re-Pair on a string of length n in $\mathcal{O}(n^2)$ time with $\max((n/c) \lg n,$ $n \lceil \lg \tau \rceil) + \mathcal{O}(\lg n)$ bits of working space including the text space as a rewritable part in the working space, where $c \geq 1$ is a fixed constant and $\tau = \sigma_m$ is the sum of the alphabet size σ and the number of non-terminal symbols.*

In our model, we assume that we can enlarge the text T_i from $n_i \lceil \lg \sigma_i \rceil$ bits to $n_i \lceil \lg \sigma_{i+1} \rceil$ bits without additional extra memory. Our main idea is to store a growing frequency table using the space freed up by replacing bigrams with non-terminals. In detail, we maintain a frequency table F in T_i of length f_k for a growing variable f_k, which is set to $f_0 := \mathcal{O}(1)$ in the beginning. The table F takes $f_k \lceil \lg(\sigma_i^2 n/2) \rceil$ bits, which is $\mathcal{O}(\lg(\sigma^2 n)) = \mathcal{O}(\lg n)$ bits for $k = 0$. When we want to query it for a most frequent bigram, we linearly scan F in $\mathcal{O}(f_k) = \mathcal{O}(n)$ time, which is not a problem since (a) the number of queries is $m \leq n$ and (b) we aim for $\mathcal{O}(n^2)$ as the overall running time. A consequence is that there is no need to sort the bigrams in F according to their frequencies, which simplifies the following discussion.

Frequency table F: With Lemma 2, we can compute F in $\mathcal{O}(n \max(n, f_k) \lg f_k / f_k)$ time. Instead of recomputing F on every turn i, we want to recompute it only when it no longer stores a most frequent bigram. However, it is not obvious when this happens as replacing a most frequent bigram during a turn (a) removes this entry in F and (b) can reduce the frequencies of other bigrams in F, making them possibly less frequent than other bigrams not tracked by F. Hence, the variable i for the i-th turn (creating the i-th non-terminal) and the variable k for recomputing the frequency table F the $(k+1)$-st time are loosely connected. We group together all turns with the same f_k and call this group the *k-th round* of the algorithm. At the beginning of each round, we enlarge f_k and create a new F with a capacity for f_k bigrams. Since a recomputation of F takes much time, we want to end a round only if F is no longer useful, i.e., when we no longer can guarantee that F stores a most frequent bigram. To achieve our claimed time bounds, we want to assign all m turns to $\mathcal{O}(\lg n)$ different rounds, which can only be done if f_k grows sufficiently fast.

Algorithm outline: At the beginning of the k-th round and the i-th turn, we compute the frequency table F storing f_k bigrams and keep additionally the lowest frequency of F as a threshold t_k, which is treated as a constant during this round. During the computation of the i-th turn, we replace the most frequent bigram (say, $bc \in \Sigma_i^2$) in the text T_i with a non-terminal X_{i+1} to produce T_{i+1}. Thereafter, we remove bc from F and update those frequencies in F, which were decreased by the replacement of bc with X_{i+1} and add each bigram containing the new character X_{i+1} into F if its frequency is at least t_k. Whenever a frequency in F drops below t_k, we discard it. If F becomes empty, we move to the $(k+1)$-st round and create a new F for storing f_{k+1} frequencies. Otherwise (F still stores an entry), we can be sure that F stores a most frequent bigram. In both cases, we recurse with the $(i+1)$-st turn by selecting the bigram with the highest frequency stored in F. We show in Algorithm 1 the pseudo code of this outlined algorithm. We describe in the following how we update F and how large f_{k+1} can become at least.

2.3. Algorithmic Details

Suppose that we are in the k-th round and in the i-th turn. Let t_k be the lowest frequency in F computed at the beginning of the k-th round. We keep t_k as a constant threshold for the invariant that all frequencies in F are at least t_k during the k-th round. With this threshold, we can assure in the following that F is either empty or stores a most frequent bigram.

Now suppose that the most frequent bigram of T_i is $bc \in \Sigma_i^2$, which is stored in F. To produce T_{i+1} (and hence advancing to the $(i+1)$-st turn), we enlarge the space of T_i from $n_i \lceil \lg \sigma_i \rceil$ to $n_i \lceil \lg \sigma_{i+1} \rceil$ and replace all occurrences of bc in T_i with a new non-terminal X_{i+1}. Subsequently, we would like to take the next bigram of F. For that, however, we

---

Algorithm 1: Algorithmic outline of our proposed algorithm working on a text T with a growing frequency table F. The constants α_i and β_i are explained in Section 2.3. The same section shows that the outer while loop is executed $\mathcal{O}(\lg n)$ times.

---

1   $k \leftarrow 0, i \leftarrow 0$
2   $f_0 \leftarrow \mathcal{O}(1)$
3   $T_0 \leftarrow T$
4   **while** *highest frequency of a bigram in T is greater than one* **do** $\triangleright$ during the k-th round
5     $F \leftarrow$ frequency table of Lemma 2 with $d := f_k$
6     $t_k \leftarrow$ minimum frequency stored in F
7   **while** $F \neq \varnothing$ **do** $\triangleright$ during the i-th turn
8   bc $\leftarrow$ most frequent bigram stored in F
9       $T_{i+1} \leftarrow T_i.\text{replace}(\text{bc}, X_{i+1})$ $\triangleright$ create rule $X_{i+1} \to$ bc
10      $i \leftarrow i + 1$ $\triangleright$ introduce the $(i+1)$-th turn
11   remove all bigrams with frequency lower than t_k from F
12   add new bigrams to F having X_i as left or right character and a frequency of at least t_k
13     $f_{k+1} \leftarrow f_k + \max(2/\beta_i, (f_k - 1)/(2\beta_i))/\alpha_i$
14     $k \leftarrow k + 1$ $\triangleright$ introduce the $(k+1)$-th round
15   Invariant: $i = m$ (the number of non-terminals)

---

need to update the stored frequencies in F. To see this necessity, suppose that there is an occurrence of abcd with two characters $a, d \in \Sigma_i$ in T_i. By replacing bc with X_{i+1},

1. the frequencies of ab and cd decrease by one (for the border case a = b = c (resp. b = c = d), there is no need to decrement the frequency of ab (resp. cd)), and
2. the frequencies of aX_{i+1} and $X_{i+1}d$ increase by one.

Updating F. We can take care of the former changes (1) by decreasing the respective bigram in F (in the case that it is present). If the frequency of this bigram drops below the threshold t_k, we remove it from F as there may be bigrams with a higher frequency that are not present in F. To cope with the latter changes (2), we track the characters adjacent to X_{i+1} after the replacement, count their numbers, and add their respective bigrams to F if their frequencies are sufficiently high. In detail, suppose that we have substituted bc with X_{i+1} exactly h times. Consequently, with the new text T_{i+1} we have additionally $h \lg \sigma_{i+1}$ bits of free space (the free space is consecutive after shifting all characters to the left), which we call D in the following. Subsequently, we scan the text and put the characters of Σ_{i+1} appearing to the left of each of the h occurrences of X_{i+1} into D. After sorting the characters in D lexicographically, we can count the frequency of aX_{i+1} for each character $a \in \Sigma_{i+1}$ preceding an occurrence of X_{i+1} in the text T_{i+1} by scanning D linearly. If the obtained frequency of such a bigram aX_{i+1} is at least as high as the threshold t_k, we insert aX_{i+1} into F and subsequently discard a bigram with the currently lowest frequency in F if the size of F has become $f_k + 1$. In the case that we visit a run of X_{i+1}'s during the creation of D, we must take care of not counting the overlapping occurrences of $X_{i+1}X_{i+1}$. Finally, we can count analogously the occurrences of $X_{i+1}d$ for all characters $d \in \Sigma_i$ succeeding an occurrence of X_{i+1}.

Capacity of F: After the above procedure, we update the frequencies of F. When F becomes empty, all bigrams stored in F are replaced or have a frequency that becomes less than t_k. Subsequently, we end the k-th round and continue with the $(k + 1)$-st round by (a) creating a new frequency table F with capacity f_{k+1} and (b) setting the new threshold t_{k+1} to the minimal frequency in F. In what follows, we (a) analyze in detail when F becomes empty (as this determines the sizes f_k and f_{k+1}) and (b) show that we can compensate the number of discarded bigrams with an enlargement of F's capacity from f_k bigrams to f_{k+1} bigrams for the sake of our aimed total running time.

Next, we analyze how many characters we have to free up (i.e., how many bigram occurrences we have to replace) to gain enough space for storing an additional frequency. Let $\delta_i := \lg(\sigma_{i+1}^2 n_i/2)$ be the number of bits needed to store one entry in F, and let $\beta_i := \min(\delta_i/\lg\sigma_{i+1}, c\delta_i/\lg n)$ be the minimum number of characters that need to be freed to store one frequency in this space. To understand the value of β_i, we look at the arguments of the minimum function in the definition of β_i and simultaneously at the maximum function in our aimed working space of $\max(n\lceil\lg\sigma_m\rceil, (n/c)\lg n) + \mathcal{O}(\lg n)$ bits (cf. Theorem 1):

1. The first item in this maximum function allows us to spend $\lg\sigma_{i+1}$ bits for each freed character such that we obtain space for one additional entry in F after freeing $\delta_i/\lg\sigma_{i+1}$ characters.
2. The second item allows us to use $\lg n$ additional bits after freeing up c characters. This additional treatment helps us to let f_k grow sufficiently fast in the first steps to save our $\mathcal{O}(n^2)$ time bound, as for sufficiently small alphabets and large text sizes, $\lg(\sigma^2 n/2)/\lg\sigma = \mathcal{O}(\lg n)$, which means that we might run the first $\mathcal{O}(\lg n)$ turns with $f_k = \mathcal{O}(1)$ and, therefore, already spend $\mathcal{O}(n^2\lg n)$ time. Hence, after freeing up $c\delta_i/\lg n$ characters, we have space to store one additional entry in F.

With $\beta_i = \min(\delta_i/\lg\sigma_{i+1}, c\delta_i/\lg n) = \mathcal{O}(\log_\sigma n) \cap \mathcal{O}(\log_n \sigma) = \mathcal{O}(1)$, we have the sufficient condition that replacing a constant number of characters gives us enough space for storing an additional frequency.

If we assume that replacing the occurrences of a bigram stored in F does not decrease the other frequencies stored in F, the analysis is now simple: Since each bigram in F has a frequency of at least two, $f_{k+1} \geq f_k + f_k/\beta_i$. Since $\beta_i = \mathcal{O}(1)$, this lets f_k grow exponentially, meaning that we need $\mathcal{O}(\lg n)$ rounds. In what follows, we show that this is also true in the general case.

Lemma 3. *Given that the frequency of all bigrams in F drops below the threshold t_k after replacing the most frequent bigram $\mathtt{bc}$, then its frequency has to be at least $\max(2, |F| - 1/2)$, where $|F| \leq f_k$ is the number of frequencies stored in F.*

Proof. If the frequency of bc in T_i is x, then we can reduce at most $2x$ frequencies of other bigrams (both the left character and the right character of each occurrence of bc can contribute to an occurrence of another bigram). Since a bigram must occur at least twice in T_i to be present in F, the frequency of bc has to be at least $\max(2, (f_k - 1)/2)$ for discarding all bigrams of F. $\square$

Suppose that we have enough space available for storing the frequencies of $\alpha_i f_k$ bigrams, where α_i is a constant (depending on σ_i and n_i) such that F and the working space of Lemma 2 with $d = f_k$ can be stored within this space. With β_i and Lemma 3 with $|F| = f_k$, we have:

$$\begin{aligned}
\alpha_i f_{k+1} &= \alpha_i f_k + \max(2/\beta_i, (f_k - 1)/(2\beta_i)) \\
&= \alpha_i f_k \max(1 + 2/(\alpha_i\beta_i f_k), 1 + 1/(2\alpha_i\beta_i) - 1/(2\alpha_i\beta_i f_k)) \\
&\geq \alpha_i f_k(1 + 2/(5\alpha_i\beta_i)) =: \gamma_i \alpha_i f_k \text{ with } \gamma_i := 1 + 2/(5\alpha_i\beta_i),
\end{aligned}$$

where we use the equivalence $1 + 2/(\alpha_i\beta_i f_k) = 1 + 1/(2\alpha_i\beta_i) - 1/(2\alpha_i\beta_i f_k) \Leftrightarrow 5 = f_k$ to estimate the two arguments of the maximum function.

Since we let f_k grow by a factor of at least $\gamma := \min_{1\leq i\leq m}\gamma_i > 1$ for each recomputation of F, $f_k = \Omega(\gamma^k)$, and therefore, $f_k = \Theta(n)$ after $k = \mathcal{O}(\lg n)$ steps. Consequently, after reaching $k = \mathcal{O}(\lg n)$, we can iterate the above procedure a constant number of times to compute the non-terminals of the remaining bigrams occurring at least twice.

Time analysis: In total, we have $\mathcal{O}(\lg n)$ rounds. At the start of the k-th round, we compute F with the algorithm of Lemma 2 with $d = f_k$ on a text of length at most $n - f_k$ in $\mathcal{O}(n(n - f_k) \cdot \lg f_k / f_k)$ time with $f_k \leq n$. Summing this up, we get:

$$\mathcal{O}\left(\sum_{k=0}^{\mathcal{O}(\lg n)} \frac{n - f_k}{f_k} n \lg f_k \right) = \mathcal{O}\left(n^2 \sum_k \frac{k}{\gamma^k} \right) = \mathcal{O}\left(n^2 \right) \text{ time.} \tag{1}$$

In the i-th turn, we update F by decreasing the frequencies of the bigrams affected by the substitution of the most frequent bigram bc with X_{i+1}. For decreasing such a frequency, we look up its respective bigram with a linear scan in F, which takes $f_k = \mathcal{O}(n)$ time. However, since this decrease is accompanied with a replacement of an occurrence of bc, we obtain $\mathcal{O}(n^2)$ total time by charging each text position with $\mathcal{O}(n)$ time for a linear search in F. With the same argument, we can bound the total time for sorting the characters in D to $\mathcal{O}(n^2)$ overall time: Since we spend $\mathcal{O}(h \lg h)$ time on sorting h characters preceding or succeeding a replaced character and $\mathcal{O}(f_k) = \mathcal{O}(n)$ time on swapping a sufficiently large new bigram composed of X_{i+1} and a character of Σ_{i+1} with a bigram with the lowest frequency in F, we charge each text position again with $\mathcal{O}(n)$ time. Putting all time bounds together gives the claim of Theorem 1.

2.4. Storing the Output In-Place

Finally, we show that we can store the computed grammar in text space. More precisely, we want to store the grammar in an auxiliary array A packed at the end of the working space such that the entry $A[i]$ stores the right-hand side of the non-terminal X_i, which is a bigram. Thus, the non-terminals are represented implicitly as indices of the array A. We therefore need to subtract $2 \lg \sigma_i$ bits of space from our working space $\alpha_i f_k$ after the i-th turn. By adjusting α_i in the above equations, we can deal with this additional space requirement as long as the frequencies of the replaced bigrams are at least three (we charge two occurrences for growing the space of A).

When only bigrams with frequencies of at most two remain, we switch to a simpler algorithm, discarding the idea of maintaining the frequency table F: Suppose that we work with the text T_i. Let λ be a text position, which is one in the beginning, but will be incremented in the following turns while holding the invariant $T[1..\lambda]$ that does not contain a bigram of frequency two. We scan $T_i[\lambda..n]$ linearly from left to right and check, for each text position j, whether the bigram $T_i[j]T_i[j+1]$ has another occurrence $T_i[j']T_i[j'+1] = T_i[j]T_i[j+1]$ with $j' > j + 1$, and if so,

(a)   append $T_i[j]T_i[j+1]$ to A,
(b)   replace $T_i[j]T_i[j+1]$ and $T_i[j']T_i[j'+1]$ with a new non-terminal X_{i+1} to transform T_i to T_{i+1}, and
(c)   recurse on T_{i+1} with $\lambda := j$ until no bigram with frequency two is left.

The position λ, which we never decrement, helps us to skip over all text positions starting with bigrams with a frequency of one. Thus, the algorithm spends $\mathcal{O}(n)$ time for each such text position and $\mathcal{O}(n)$ time for each bigram with frequency two. Since there are at most n such bigrams, the overall running time of this algorithm is $\mathcal{O}(n^2)$.

Remark 1 (Pointer machine model). *Refraining from the usage of complicated algorithms, our algorithm consists only of elementary sorting and scanning steps. This allows us to run our algorithm on a pointer machine, obtaining the same time bound of $\mathcal{O}(n^2)$. For the space bounds, we assume that the text is given in n words, where a word is large enough to store an element of Σ_m or a text position.*

2.5. Step-by-Step Execution

Here, we present an exemplary execution of the first turn (of the first round) on the input $T = \text{cabaacabcabaacaaabcab}$. We visualize each step of this turn as a row in Figure 1. A detailed description of each row follows:

Row 1: Suppose that we have computed F, which has the constant number of entries $f_0 = 3$ (in the later turns when the size f_k becomes larger, F will be put in the text space). The highest frequency is five achieved by ab and ca. The lowest frequency represented in F is three, which becomes the threshold t_0 for a bigram to be present in F such that bigrams whose frequencies drop below t_0 are removed from F. This threshold is a constant for all later turns until F is rebuilt (in the following round). During Turn 1, the algorithm proceeds now as follows:

Row 2: Choose ab as a bigram to replace with a new non-terminal X_1 (break ties arbitrarily). Replace every occurrence of ab with X_1 while decrementing frequencies in F according to the neighboring characters of the replaced occurrence.

Row 3: Remove from F every bigram whose frequency falls below the threshold. Obtain space for D by aligning the compressed text T_1 (the process of Row 2 and Row 3 can be done simultaneously).

Row 4: Scan the text and copy each character preceding an occurrence of X_1 in T_1 to D.

Row 5: Sort characters in D lexicographically.

Row 6: Insert new bigrams (consisting of a character of D and X_1) whose frequencies are at least as large as the threshold.

Row 7: Scan the text again and copy each character succeeding an occurrence of X_1 in T_1 to D (symmetric to Row 4).

Row 8: Sort all characters in D lexicographically (symmetric to Row 5).

Row 9: Insert new bigrams whose frequencies are at least as large as the threshold (symmetric to Row 6).

c	a	b	a	a	c	a	b	c	a	b	a	a	c	a	a	a	b	c	a	b	ab	ca	aa
c	1		a	a	c	1		c	1		a	a	c	a	a	1		c	1		ab	ca	aa

c	1	a	a	c	1	c	1	a	a	c	a	a	1	c	1								aa
c	1	a	a	c	1	c	1	a	a	c	a	a	1	c	1	c	c	c	a	c			aa
c	1	a	a	c	1	c	1	a	a	c	a	a	1	c	1	a	c	c	c	c			aa
c	1	a	a	c	1	c	1	a	a	c	a	a	1	c	1						c	1	aa
c	1	a	a	c	1	c	1	a	a	c	a	a	1	c	1	a	c	a	c		c	1	aa
c	1	a	a	c	1	c	1	a	a	c	a	a	1	c	1	a	a	c	c		c	1	aa
c	1	a	a	c	1	c	1	a	a	c	a	a	1	c	1						c	1	aa

Figure 1. Step-by-step execution of the first turn of our algorithm on the string $T = \text{cabaacabcabaacaaabcab}$. The turn starts with the memory configuration given in Row 1. Positions 1 to 21 are text positions, and Positions 22 to 24 belong to F ($f_0 = 3$, and it is assumed that a frequency fits into a text entry). Subsequent rows depict the memory configuration during Turn 1. A comment on each row is given in Section 2.5.

2.6. Implementation

At https://github.com/koeppl/repair-inplace, we provide a simplified implementation in C++17. The simplification is that we (a) fix the bit width of the text space to 16 bit and (b) assume that Σ is the byte alphabet. We further skip the step increasing the bit width of the text from $\lg \sigma_i$ to $\lg \sigma_{i+1}$. This means that the program works as long as the characters of Σ_m fit into 16 bits. The benchmark, whose results are displayed in Table 1, was conducted on a Mac Pro Server with an Intel Xeon CPU X5670 clocked at 2.93 GHz running Arch Linux. The implementation was compiled with gcc-8.2.1 in the highest optimization mode -O3. Looking at Table 1, we observe that the running time is super-linear to the input size on all text instances, which we obtained from the Pizza&Chili corpus (http://pizzachili.dcc.uchile.cl/). We conducted the same experiments with an implementation of Gonzalo Navarro (https://users.dcc.uchile.cl/~gnavarro/software/repair.tgz) in Table 2 with considerably better running times while restricting the algorithm to use 1 MB of RAM during execution. Table 3 gives some characteristics about the used data sets. We see that the number of rounds is the number of turns plus one for every unary string a^{2^k} with an integer $k \geq 1$ since the text contains only one bigram with a frequency larger than two in each round. Replacing this bigram in the text makes F empty such that the algorithm recomputes F after each turn. Note that the number of rounds can drop while scaling the prefix length based on the choice of the bigrams stored in F.

Table 1. Experimental evaluation of our implementation and the implementation of Navarro described in Section 2.6. Table entries are running times in seconds. The last line is the benchmark on the unary string $aa \cdots a$.

Data Set	Our Implementation					Implementation of Navarro				
	Prefix Size in KiB									
	64	128	256	512	1024	64	128	256	512	1024
ESCHERICHIA_COLI	20.68	130.47	516.67	1708.02	10,112.47	0.01	0.02	0.07	0.18	0.29
CERE	13.69	90.83	443.17	2125.17	9185.58	0.01	0.02	0.04	0.16	0.22
COREUTILS	12.88	75.64	325.51	1502.89	5144.18	0.01	0.05	0.05	0.14	0.29
EINSTEIN.DE.TXT	19.55	88.34	181.84	805.81	4559.79	0.01	0.04	0.08	0.10	0.25
EINSTEIN.EN.TXT	21.11	78.57	160.41	900.79	4353.81	0.01	0.02	0.05	0.21	0.51
INFLUENZA	41.01	160.68	667.58	2630.65	10,526.23	0.03	0.02	0.05	0.11	0.36
KERNEL	20.53	101.84	208.08	1575.48	5067.80	0.01	0.04	0.09	0.18	0.27
PARA	20.90	175.93	370.72	2826.76	9462.74	0.01	0.01	0.08	0.12	0.35
WORLD_LEADERS	11.92	21.82	167.52	661.52	1718.36	0.01	0.01	0.06	0.11	0.25
$aa \cdots a$	0.35	0.92	3.90	14.16	61.74	0.01	0.01	0.05	0.05	0.12

Table 2. Experimental evaluation of the implementation of Navarro. Table entries are running times in seconds.

Data Set	Prefix Size in KiB				
	64	128	256	512	1024
ESCHERICHIA_COLI	0.01	0.02	0.07	0.18	0.29
CERE	0.01	0.02	0.04	0.16	0.22
COREUTILS	0.01	0.05	0.05	0.14	0.29
EINSTEIN.DE.TXT	0.01	0.04	0.08	0.10	0.25
EINSTEIN.EN.TXT	0.01	0.02	0.05	0.21	0.51
INFLUENZA	0.03	0.02	0.05	0.11	0.36
KERNEL	0.01	0.04	0.09	0.18	0.27
PARA	0.01	0.01	0.08	0.12	0.35
WORLD_LEADERS	0.01	0.01	0.06	0.11	0.25

Table 3. Characteristics of our data sets used in Section 2.6. The number of turns and rounds are given for each of the prefix sizes 128, 256, 512, and 1024 KiB of the respective data sets. The number of turns reflecting the number of non-terminals is given in units of thousands. The turns of the unary string aa $\cdots$ a are in plain units (not divided by thousand).

| Data Set | σ | Turns/1000 Prefix Size in KiB | | | | | Rounds Prefix Size in KiB | | | | |
		2^6	2^7	2^8	2^9	2^{10}	2^6	2^7	2^8	2^9	2^{10}
ESCHERICHIA_COLI	4	1.8	3.2	5.6	10.3	18.1	6	9	9	12	12
CERE	5	1.4	2.8	5.0	9.2	15.1	13	14	14	14	14
COREUTILS	113	4.7	6.7	10.2	16.1	26.5	15	15	15	14	14
EINSTEIN.DE.TXT	95	1.7	2.8	3.7	5.2	9.7	14	14	15	16	16
EINSTEIN.EN.TXT	87	3.3	3.5	3.8	4.5	8.6	16	15	15	15	17
INFLUENZA	7	2.5	3.7	9.5	13.4	22.1	11	12	14	13	15
KERNEL	160	4.5	8.0	13.9	24.5	43.7	10	11	14	14	13
PARA	5	1.8	3.2	5.8	10.1	17.6	12	12	13	13	14
WORLD_LEADERS	87	2.6	4.3	6.1	10.0	42.1	11	11	11	11	14
aa $\cdots$ a	1	15	16	17	18	19	16	17	18	19	20

3. Bit-Parallel Algorithm

In the case that $\tau = \sigma_m$ is $o(n)$ (and therefore, $\sigma = o(n)$), a word-packing approach becomes interesting. We present techniques speeding up the previously introduced operations on chunks of $\mathcal{O}(\log_\tau n)$ characters from $\mathcal{O}(\log_\tau n)$ time to $\mathcal{O}(\lg \lg \lg n)$ time. In the end, these techniques allow us to speed up the sequential algorithm of Theorem 1 from $\mathcal{O}(n^2)$ time to the following:

Theorem 2. *We can compute Re-Pair on a string of length n in $\mathcal{O}(n^2 \lg \log_\tau n \lg \lg \lg n / \log_\tau n)$ time with $\max((n/c) \lg n, n\lceil \lg \tau \rceil) + \mathcal{O}(\lg n)$ bits of working space including the text space, where $c \geq 1$ is a fixed constant and $\tau = \sigma_m$ is the sum of the alphabet size σ and the number of non-terminal symbols.*

Note that the $\mathcal{O}(\lg \lg \lg n)$ time factor is due to the popcount function [34] (Algorithm 1), which has been optimized to a single instruction on modern computer architectures. Our toolbox consists of several elementary instructions shown in Figure 2. There, $msb(X)$ can be computed in constant time algorithm using $\mathcal{O}(\lg n)$ bits [35] (Section 5). The last two functions in Figure 2 are explained in Figure 3.

3.1. Broadword Search

First, we deal with accelerating the computation of the frequency of a bigram in T by exploiting broadword search thanks to the word RAM model. We start with the search of single characters and subsequently extend this result to bigrams:

Operation	Description
$X \ll j$	shift X j positions to the left
$X \gg j$	shift X j positions to the right
$\neg X$	bitwise NOT of X
$X \otimes Y$	bitwise XOR of X and Y
-1	bit vector consisting only of one bits
$msb(X)$	returns the position of the most significant set bit of X, i.e., $\lfloor \lg X \rfloor + 1$;
$rmPreRun(X)$	sets all bits of the maximal prefix of consecutive ones to zero
$rmSufRun(X)$	sets all bits of the maximal suffix of consecutive ones to zero

Figure 2. Operations used in Figures 4 and 5 for two bit vectors X and Y. All operations can be computed in constant time. See Figure 3 for an example of rmSufRun and rmPreRun.

rmPreRun(X)		rmSufRun(X)	
Operation	Example	Operation	Example
X	11100110	X	01100111
$\neg X$	00011001	$\neg X$	10011000
$1 \ll (1 + \mathrm{msb}(\neg X))$	00100000	$\neg X - 1$	10010111
$(1 \ll (1 + \mathrm{msb}(\neg X))) - 1$	00011111	$(\neg X - 1)\ \&\ X$	00000111
$((1 \ll (1 + \mathrm{msb}(\neg X))) - 1)\ \&\ X$	00000110	$\neg((\neg X - 1)\ \&\ X)$	11111000
		$\neg((\neg X - 1)\ \&\ X)\ \&\ X$	01100000

Figure 3. Step-by-step execution of rmPreRun(X) and rmSufRun(X) introduced in Figure 2 on a bit vector X.

Lemma 4. *We can count the occurrences of a character $c \in \Sigma$ in a string of length $\mathcal{O}(\log_\sigma n)$ in $\mathcal{O}(\lg \lg \lg n)$ time.*

Proof. Let q be the largest multiple of $\lceil \lg \sigma \rceil$ fitting into a computer word, divided by $\lceil \lg \sigma \rceil$. Let $S \in \Sigma^*$ be a string of length q. Our first task is to compute a bit mask of length $q\lceil \lg \sigma \rceil$ marking the occurrences of a character $c \in \Sigma$ in S with a '1'. For that, we follow the constant time broadword pattern matching of Knuth [36] (Section 7.1.3); see https://github.com/koeppl/broadwordsearch for a practical implementation. Let H and L be two bit vectors of length $\lceil \lg \sigma \rceil$ having marked only the most significant or the least significant bit, respectively. Let H^q and L^q denote the q times concatenation of H and L, respectively. Then, the operations in Figure 4 yield an array X of length q with:

$$X[i] = \begin{cases} 2^{\lceil \lg \sigma \rceil} - 1 & \text{if } S[i] = c, \\ 0 & \text{otherwise,} \end{cases} \tag{2}$$

where each entry of X has $\lceil \lg \sigma \rceil$ bits.

Operation	Description		Example
read S			$101010000 \to S$
$X \leftarrow S \otimes c^q$	match S with c^q; $X[i] = 0 \Leftrightarrow$ $S[i] = c$	$\otimes$	$101010000 = S$ 010010010
		$=$	$111000010 \to X$
$Y \leftarrow X - L^q$		$-$	$111000010 = X$ 001001001
		$=$	$101111001 \to Y$
$X \leftarrow Y\ \&\ \neg X$	$X[i]\ \&\ 2^{\lceil \lg \sigma \rceil} - 1\ =\ 1 \Leftrightarrow$ $S[i] = c$	$\&$	$101111001 = Y$ 000111101
		$=$	$000111001 \to X$
$X \leftarrow X\ \&\ H^q$	$X[i] = 0 \Leftrightarrow S[i] \neq c$	$\&$	$000111001 = X$ 100100100
		$=$	$000100000 \to X$
$X \leftarrow (X - (X \gg (\lceil \lg \sigma \rceil - 1))) \mid X$	X as in Equation (2)	$-$	$000100000 = X$ 000001000
		$=$	000011000
		$\mid$	000100000
		$=$	$000111000 \to X$

Figure 4. Matching all occurrences of a character in a string S fitting into a computer word in constant time by using bit-parallel instructions. For the last step, special care has to be taken when the last character of S is a match, as shifting X $\lceil \lg \sigma \rceil$ bits to the right might erase a '1' bit witnessing the rightmost match. In the description column, X is treated as an array of integers with bit width $\lceil \lg \sigma \rceil$. In this example, $S = 101010000$, c has the bit representation 010 with $\lg \sigma = 3$, and $q = 3$.

To obtain the number of occurrences of c in S, we use the popcount operation returning the number of zero bits in X and divide the result by $\lceil \lg \sigma \rceil$. The popcount instruction takes $\mathcal{O}(\lg \lg \lg n)$ time ([34] Algorithm 1). $\square$

Having Lemma 4, we show that we can compute the frequency of a bigram in T in $\mathcal{O}(n \lg \lg \lg n / \log_\sigma n)$ time. For that, we interpret $T \in \Sigma^n$ of length n as a text $T \in (\Sigma^2)^{\lceil n/2 \rceil}$ of length $\lceil n/2 \rceil$. Then, we partition T into strings fitting into a computer word and call each string of this partition a *chunk*. For each chunk, we can apply Lemma 4 by treating a bigram $c \in \Sigma^2$ as a single character. The result is, however, not the frequency of the bigram c in general. For computing the frequency a bigram $\mathtt{bc} \in \Sigma^2$, we distinguish the cases $\mathtt{b} \neq \mathtt{c}$ and $\mathtt{b} = \mathtt{c}$.

Case $\mathtt{b} \neq \mathtt{c}$: By applying Lemma 4 to find the character $\mathtt{bc} \in \Sigma^2$ in a chunk S (interpreted as a string of length $\lfloor q/2 \rfloor$ on the alphabet Σ^2), we obtain the number of occurrences of $\mathtt{bc}$ starting at odd positions in S. To obtain this number for all even positions, we apply the procedure to $\mathtt{d}S$ with $\mathtt{d} \in \Sigma \setminus \{\mathtt{b}, \mathtt{c}\}$. Additional care has to be taken at the borders of each chunk matching the last character of the current chunk and the first character of the subsequent chunk with $\mathtt{b}$ and $\mathtt{c}$, respectively.

Case $\mathtt{b} = \mathtt{c}$: This case is more involving as overlapping occurrences of $\mathtt{bb}$ can occur in S, which we must not count. To this end, we watch out for *runs* of $\mathtt{b}$'s, i.e., substrings of maximal lengths consisting of the character $\mathtt{b}$ (here, we consider also maximal substrings of $\mathtt{b}$ with length one as a run). We separate these runs into runs ending either at even or at odd positions. We do this because the frequency of $\mathtt{bb}$ in a run of $\mathtt{b}$'s ending at an even (resp. odd) position is the number of occurrences of $\mathtt{bb}$ within this run ending at an even (resp. odd) position. We can compute these positions similarly to the approach for $\mathtt{b} \neq \mathtt{c}$ by first (a) hiding runs ending at even (resp. odd) positions and then (b) counting all bigrams ending at even (resp. odd) positions. Runs of $\mathtt{b}$'s that are a prefix or a suffix of S are handled individually if S is neither the first nor the last chunk of T, respectively. That is because a run passing a chunk border starts and ends in different chunks. To take care of those runs, we remember the number of $\mathtt{b}$'s of the longest suffix of every chunk and accumulate this number until we find the end of this run, which is a prefix of a subsequent chunk. The procedure for counting the frequency of $\mathtt{bb}$ inside S is explained with an example in Figure 5. With the aforementioned analysis of the runs crossing chunk borders, we can extend this procedure to count the frequency of $\mathtt{bb}$ in T. We conclude:

Lemma 5. *We can compute the frequency of a bigram in a string T of length n whose characters are drawn from an alphabet of size σ in $\mathcal{O}(n \lg \lg \lg n / \log_\sigma n)$ time.*

3.2. Bit-Parallel Adaption

Similarly to Lemma 2, we present an algorithm computing the d most frequent bigrams, but now with the word-packed search of Lemma 5.

Lemma 6. *Given an integer d with $d \geq 1$, we can compute the frequencies of the d most frequent bigrams in a text of length n whose characters are drawn from an alphabet of size σ in $\mathcal{O}(n^2 \lg \lg \lg n / \log_\sigma n)$ time using $d \lceil \lg(\sigma^2 n/2) \rceil + \mathcal{O}(\lg n)$ bits.*

Proof. We allocate a frequency table F of length d. For each text position i with $1 \leq i \leq n - 1$, we compute the frequency of $T[i]T[i+1]$ in $\mathcal{O}(n \lg \lg \lg n / \log_\sigma n)$ time with Lemma 5. After computing a frequency, we insert it into F if it is one of the d most frequent bigrams among the bigrams we have already computed. We can perform the insertion in $\mathcal{O}(\lg d)$ time if we sort the entries of F by their frequencies, obtaining $\mathcal{O}((n \lg \lg \lg n / \log_\sigma n + \lg d)n)$ total time. $\square$

Operation	Description	Example
input S		bbdbbdcbbbdbb $= S$
$X \leftarrow \text{find}(\text{b}, S)$	search b in S	$1101100111011 \rightarrow X$
$X \leftarrow \text{rmPreRun}(X)$	erase prefix of b's	$0001100111011 \rightarrow X$
$M \leftarrow \text{rmSufRun}(X)$	erase suffix of b's	$0001100111000 \rightarrow M$
$B \leftarrow \text{findBigram}(01, M) \,\&\, M$	starting of each b run	$0001000100000 \rightarrow B$
$E \leftarrow \text{findBigram}(10, M) \,\&\, M$	end of each b run	$0000100001000 \rightarrow E$
$M \leftarrow M \,\&\, \neg B$	trim head of runs	$0000100011000 \rightarrow M$
$X \leftarrow B - (E \,\&\, (01)^{q/2})$	bit mask for all runs ending at even positions	$\begin{aligned} &0001000100000 = B \\ -\;&(0000100001000\,\& \\ &0101010101010) \\ =\;&0001000011000 \rightarrow X \end{aligned}$
$X \leftarrow M \,\&\, X$	occurrences of all bs belonging to runs ending at even positions	$\begin{aligned} &0001000011000 = X \\ \&\;&0000100011000 = M \\ =\;&0000000011000 \rightarrow X \end{aligned}$
$\text{popcount}(X \,\&\, (01)^{q/2})$	frequency of all bbs belonging to runs ending at even positions	$\begin{aligned} &0000000011000 = X \\ \&\;&0101010101010 \\ =\;&0000000001000 \end{aligned}$
$X \leftarrow B - (E \,\&\, (10)^{q/2})$	bit mask for all runs ending at odd positions	$\begin{aligned} &0001000100000 = B \\ -\;&(0000100001000\,\& \\ &1010101010101) \\ =\;&0000100100000 \rightarrow X \end{aligned}$
$X \leftarrow M \,\&\, X$	occurrences of all bs belonging to runs ending at odd positions	$\begin{aligned} &0000100100000 = X \\ \&\;&0000100011000 = M \\ =\;&0000100000000 \rightarrow X \end{aligned}$
$\text{popcount}(X \,\&\, (10)^{q/2})$	frequency of all bbs belonging to runs ending at odd positions	$\begin{aligned} &0000100000000 = X \\ \&\;&1010101010101 \\ =\;&0000100000000 \end{aligned}$

Figure 5. Finding a bigram bb in a string S of bit length q, where q is the largest multiple of $2\lceil \lg \sigma \rceil$ fitting into a computer word, divided by $\lceil \lg \sigma \rceil$. In the example, we represent the strings M, B, E, and X as arrays of integers with bit width $x := \lceil \lg \sigma \rceil$ and write 1 and 0 for 1^x and 0^x, respectively. Let $\text{findBigram}(\text{bc}, X) := \text{find}(\text{bc}, X) \mid \text{find}(\text{bc}, \text{d}X)$ for $\text{d} \neq \text{b}$ be the frequency of a bigram bc with $\text{b} \neq \text{c}$ as described in Section 3.1, where the function find returns the output described in Figure 4. Each of the popcount queries gives us one occurrence as a result (after dividing the returned number by $\lceil \lg \sigma \rceil$), thus the frequency of bb in S, without looking at the borders of S, is two. As a side note, modern computer architectures allow us to shrink the 0^x or 1^x blocks to single bits by instructions like _pext_u64 taking a single CPU cycle.

Studying the final time bounds of Equation (1) for the sequential algorithm of Section 2, we see that we spend $\mathcal{O}(n^2)$ time in the first turn, but spend less time in later turns. Hence, we want to run the bit-parallel algorithm only in the first few turns until f_k becomes so large that the benefits of running Lemma 2 outweigh the benefits of the bit-parallel approach of Lemma 6. In detail, for the k-th round, we set $d := f_k$ and run the algorithm of Lemma 6 on the current text if d is sufficiently small, or otherwise the algorithm of Lemma 2. In total, we obtain:

$$\mathcal{O}\left(\sum_{k=0}^{\mathcal{O}(\lg n)}\min\left(\frac{n-f_k}{f_k}n\lg f_k,\frac{(n-f_k)^2\lg\lg\lg n}{\log_\tau n}\right)\right)=\mathcal{O}\left(n^2\sum_{k=0}^{\lg n}\min\left(\frac{k}{\gamma^k},\frac{\lg\lg\lg n}{\log_\tau n}\right)\right)$$
$$=\mathcal{O}\left(\frac{n^2\lg\log_\tau n\lg\lg\lg n}{\log_\tau n}\right)\text{ time in total,}$$

(3)

where $\tau=\sigma_m$ is the sum of the alphabet size σ and the number of non-terminals, and $k/\gamma^k>\lg\lg\lg n/\log_\tau n\Leftrightarrow k=\mathcal{O}(\lg(\lg n/(\lg\tau\lg\lg\lg n)))$.

To obtain the claim of Theorem 2, it is left to show that the k-th round with the bit-parallel approach uses $\mathcal{O}(n^2\lg\lg\lg n/\log_\tau n)$ time, as we now want to charge each text position with $\mathcal{O}(n/\log_\tau n)$ time with the same amortized analysis as after Equation (1). We target $\mathcal{O}(n/\log_\tau n)$ time for:

(1)   replacing all occurrences of a bigram,
(2)   shifting freed up text space to the right,
(3)   finding the bigram with the highest or lowest frequency in F,
(4)   updating or exchanging an entry in F, and
(5)   looking up the frequency of a bigram in F.

Let $x:=\lceil\lg\sigma_{i+1}\rceil$ and q be the largest multiple of x fitting into a computer word, divided by x. For Item (1), we partition T into substrings of length q and apply Item (1) to each such substring S. Here, we combine the two bit vectors of Figure 5 used for the two popcount calls by a bitwise OR and call the resulting bit vector Y. Interpreting Y as an array of integers of bit width x, Y has q entries, and it holds that $Y[i]=2^x-1$ if and only if $S[i]$ is the second character of an occurrence of the bigram we want to replace. (Like in Item (1), the case in which the bigram crosses a boundary of the partition of T is handled individually). We can replace this character in all marked positions in S by a non-terminal X_{i+1} using x bits with the instruction $(S\ \&\ \neg Y)\ |\ ((Y\ \&\ L^q)\cdot X_{i+1})$, where L with $|L|=x$ is the bit vector having marked only the least significant bit. Subsequently, for Item (2), we erase all characters $S[i]$ with $Y[i+1]=(Y\ll x)[i]=2^x-1$ and move them to the right of the bit chunk S sequentially. In the subsequent bit chunks, we can use word-packed shifting. The sequential bit shift costs $\mathcal{O}(|S|)=\mathcal{O}(\log_{\sigma_{i+1}}n)$ time, but on an amortized view, a deletion of a character is done at most once per original text position.

For the remaining points, our trick is to represent F by a minimum and a maximum heap, both realized as array heaps. For the space increase, we have to lower α_i (and γ_i) adequately. Each element of an array heap stores a frequency and a pointer to a bigram stored in a separate array B storing all bigrams consecutively. A pointer array P stores pointers to the respective frequencies in both heaps for each bigram of B. The total data structure can be constructed at the beginning of the k-th round in $\mathcal{O}(f_k)$ time and hence does not worsen the time bounds. While B solves Item (5), the two heaps with P solve Items (3) and (4) even in $\mathcal{O}(\lg f_k)$ time.

In the case that we want to store the output in working space, we follow the description of Section 2.4, where we now use word-packing to find the second occurrence of a bigram in T_i in $\mathcal{O}(n/\log_{\sigma_i}n)$ time.

4. Computing MR-Re-Pair in Small Space

We can adapt our algorithm to compute the MR-Re-Pair grammar scheme proposed by Furuya et al. [18]. The difference to Re-Pair is that MR-Re-Pair replaces the most frequent maximal repeat instead of the most frequent bigram, where a maximal repeat is a reoccurring substring of the text whose frequency decreases when extending it to the left or to the right. (Here, we naturally extended the definition of *frequency* from bigrams to substrings meaning the number of non-overlapping occurrences.) Our idea is to exploit the fact that a most frequent bigram corresponds to a most frequent maximal repeat ([18], Lemma 2). This means that we can find a most frequent maximal repeat by extending all occurrences of a most frequent bigram to their left and to their right until all are no longer equal substrings. Although such an extension can be time consuming, this time is amortized by the number of characters that are replaced on creating an MR-Re-Pair rule. Hence, we conclude that we can compute MR-Re-Pair in the same space and time bounds as our algorithms (Thms. 1 and 2) computing the Re-Pair grammar.

5. Parallel Algorithm

Suppose that we have p processors on a concurrent read concurrent write (CRCW) machine, supporting in particular parallel insertions of elements and frequency updates in a frequency table. In the parallel setting, we allow us to spend $\mathcal{O}(p \lg n)$ bits of additional working space such that each processor has an extra budget of $\mathcal{O}(\lg n)$ bits. In our computational model, we assume that the text is stored in p parts of equal lengths, which we can achieve by padding up the last part with dummy characters to have n/p characters for each processor, such that we can enlarge a text stored in $n \lg \sigma$ bits to $n(\lg \sigma + 1)$ bits in $\max(1, n/p)$ time without extra memory. For our parallel variant computing Re-Pair, our working horse is a parallel sorting algorithm:

Lemma 7 ([37]). *We can sort an array of length n in $\mathcal{O}(\max(n/p, 1) \lg^2 n)$ parallel time with $\mathcal{O}(p \lg n)$ bits of working space. The work is $\mathcal{O}(n \lg^2 n)$.*

The parallel sorting allows us to state Lemma 2 in the following way:

Lemma 8. *Given an integer d with $d \geq 1$, we can compute the frequencies of the d most frequent bigrams in a text of length n whose characters are drawn from an alphabet of size σ in $\mathcal{O}(\max(n, d) \max(n/p, 1) \lg^2 d/d)$ time using $2d \lceil \lg(\sigma^2 n/2) \rceil + \mathcal{O}(p \lg n)$ bits. The work is $\mathcal{O}(\max(n, d) n \lg^2 d/d)$.*

Proof. We follow the computational steps of Lemma 2, but (a) divide a scan into p parts, (b) conduct a scan in parallel but a binary search sequentially, and (c) use Lemma 7 for the sorting. This gives us the following time bounds for each operation:

Operation	Lemma 2	Parallel
fill F' with bigrams	$\mathcal{O}(d)$	$\mathcal{O}(\max(d/p, 1))$
sort F' lexicographically	$\mathcal{O}(d \lg d)$	$\mathcal{O}(\max(d/p, 1) \lg^2 d)$
compute frequencies of F'	$\mathcal{O}(n \lg d)$	$\mathcal{O}(n/p \lg d)$
merge F' with F	$\mathcal{O}(d \lg d)$	$\mathcal{O}(\max(d/p, 1) \lg^2 d)$

The $\mathcal{O}(n/d)$ merge steps are conducted in the same way, yielding the bounds of this lemma. $\square$

In our sequential model, we produce T_{i+1} by performing a left shift of the gained space after replacing all occurrences of a most frequent bigram with a new non-terminal X_{i+1} such that we accumulate all free space at the end of the text. As described in our computational model, our text is stored as a partition of p substrings, each assigned to one processor. Instead of gathering the entire free space at T's end, we gather free space at the end of each of these substrings. We bookkeep the size and location of each such free space (there are at most p many) such that we can work on the remaining text T_{i+1} like it would be a single continuous array (and not fragmented into p substrings). This shape allows us to perform the left shift in $\mathcal{O}(n/p)$ time, while spending $\mathcal{O}(p \lg n)$ bits of space for maintaining the locations of the free space fragments.

For $p \leq n$, exchanging Lemma 2 with Lemma 8 in Equation (1) gives:

$$\mathcal{O}\left(\sum_{k=0}^{\mathcal{O}(\lg n)} \frac{n - f_k}{f_k} \frac{n}{p} \lg^2 f_k\right) = \mathcal{O}\left(\frac{n^2}{p} \sum_{k}^{\lg n} \frac{k^2}{\gamma^k}\right) = \mathcal{O}\left(\frac{n^2}{p}\right) \text{ time in total.}$$

It is left to provide an amortized analysis for updating the frequencies in F during the i-th turn. Here, we can charge each text position with $\mathcal{O}(n/p)$ time, as we have the following time bounds for each operation:

Operation	Sequential	Parallel
linearly scan F	$\mathcal{O}(f_k)$	$\mathcal{O}(f_k/p)$
linearly scan T_i	$\mathcal{O}(n_i)$	$\mathcal{O}(n_i/p)$
sort D with $h = \lvert D \rvert$	$\mathcal{O}(h \lg h)$	$\mathcal{O}(\max(1, h/p) \lg^2 h)$

The first operation in the above table is used, among others, for finding the bigram with the lowest or highest frequency in F. Computing the lowest or highest frequency in F can be done with a single variable pointing to the currently found entry with the lowest or highest frequency during a parallel scan thanks to the CRCW model. In the concurrent read exclusive write (CREW) model, concurrent writes are not possible. A common strategy lets each processor compute the entry of the lowest or highest frequency within its assigned range in F, which is then merged in a tournament tree fashion, causing $\mathcal{O}(\lg p)$ additional time.

Theorem 3. *We can compute Re-Pair in $\mathcal{O}(n^2/p)$ time with $p \leq n$ processors on a CRCW machine with $\max((n/c) \lg n, n\lceil \lg \tau \rceil) + \mathcal{O}(p \lg n)$ bits of working space including the text space, where $c \geq 1$ is a fixed constant and $\tau = \sigma_m$ is the sum of the alphabet size σ and the number of non-terminal symbols. The work is $\mathcal{O}(n^2)$.*

6. Computing Re-Pair in External Memory

This part is devoted to the first external memory (EM) algorithms computing Re-Pair, which is another way to overcome the memory limitation problem. We start with the definition of the EM model, present an approach using a sophisticated heap data structure, and another approach adapting our in-place techniques.

For the following, we use the EM model of Aggarwal and Vitter [38]. It features fast internal memory (IM) holding up to M data words and slow EM of unbounded size. The measure of the performance of an algorithm is the number of input and output operations (I/Os) required, where each I/O transfers a block of B consecutive words between memory levels. Reading or writing n contiguous words from or to disk requires $\text{scan}(n) = \Theta(n/B)$ I/Os. Sorting n contiguous words requires $\text{sort}(n) = \mathcal{O}((n/B) \cdot \log_{M/B}(n/B))$ I/Os. For realistic values of n, B, and M, we stipulate that $\text{scan}(n) < \text{sort}(n) \ll n$.

A simple approach is based on an EM heap maintaining the frequencies of all bigrams in the text. A state-of-the-art heap is due to Jiang and Larsen [39] providing insertion, deletion, and the retrieval of the maximum element in $\mathcal{O}(B^{-1} \log_{M/B}(N/B))$ I/Os, where N is the size of the heap. Since $N \leq n$, inserting all bigrams takes at most $\text{sort}(n)$ I/Os. As there are at most n additional insertions, deletions, and maximum element retrievals, this sums to at most $4 \text{sort}(n)$ I/Os. Given Re-Pair has m turns, we need to scan the text m times to replace the occurrences of all m retrieved bigrams, triggering $m \sum_{i=1}^{m} \text{scan}(\lvert T_i \rvert) \leq m \text{scan}(n)$ I/Os.

In the following, we show an EM Re-Pair algorithm that evades the use of complicated data structures and prioritizes scans over sorting. This algorithm is based on our Re-Pair algorithm. It uses Lemma 2 with $d := \Theta(M)$ such that F and F' can be kept in RAM.

This allows us to perform all sorting steps and binary searches in IM without additional I/O. We only trigger I/O operations for scanning the text, which is done $\lceil n/d \rceil$ times, since we partition T into d substrings. In total, we spend at most mn/M scans for the algorithm of Lemma 2. For the actual algorithm having m turns, we update F m times, during which we replace all occurrences of a chosen bigram in the text. This gives us m scans in total. Finally, we need to reason about D, which we create m times. However, D may be larger than M, such that we may need to store it in EM. Given that D_i is D in the i-th turn, we sort D in EM, triggering $\mathrm{sort}(D_i)$ I/Os. With the converse of Jensen's inequality ([40], Theorem B) (set there $f(x) := n \lg n$), we obtain $\sum_{i=1}^{m} \mathrm{sort}(|D_i|) \leq \mathrm{sort}(n) + \mathcal{O}(n \log_{M/B} 2)$ total I/Os for all instances of D. We finally obtain:

Theorem 4. *We can compute Re-Pair with* $\min(4 \, \mathrm{sort}(n), (mn/M) \, \mathrm{scan}(n) + \mathrm{sort}(n) + \mathcal{O}(n \log_{M/B} 2)) + m \, \mathrm{scan}(n)$ *I/Os in external memory.*

The latter approach can be practically favorable to the heap based approach if $m = o(\lg n)$ and $mn/M = o(\lg n)$, or if the EM space is also of major concern.

7. Heuristics for Practicality

The achieved quadratic or near-quadratic time bounds (Thms. 1 and 2) seem to convey the impression that this work is only of purely theoretic interest. However, we provide here some heuristics, which can help us to overcome the practical bottleneck at the beginning of the execution, where only $\mathcal{O}(\lg n)$ of bits of working space are available. In other words, we want to study several heuristics to circumvent the need to call Lemma 2 with a small parameter d, as such a case means a considerable time loss. Even a single call of Lemma 2 with a small d prevents the computation of Re-Pair of data sets larger than 1 MiB within a reasonable time frame (cf. Section 2.6). We present three heuristics depending on whether our space budget on top of the text space is within:

1. $\sigma_i^2 \lg n_i$ bits,
2. $n_i \lg(\sigma_{i+1} + n_i)$ bits, or
3. $\mathcal{O}(\lg n)$ bits.

Heuristic 1. If σ_i is small enough such that we can spend $\sigma_i^2 \lg n_i$ bits, then we can count the frequencies of all bigrams in a table of $\sigma_i^2 \lg n_i$ bits in $\mathcal{O}(n)$ time. Whenever we reach a σ_j that lets $\sigma_j \lg n_j$ grow outside of our budget, we have spent $\mathcal{O}(n)$ time in total for reaching T_j from T_i as the costs for replacements can be amortized by twice of the text length.

Heuristic 2. Suppose that we are allowed to use $(n_i - 1) \lg(n_i/2) = (n_i - 1) \lg n_i - n_i + \mathcal{O}(\lg n_i)$ bits in addition to the $n_i \lg \sigma_i$ bits of the text T_i. We create an extra array F of length $n_i - 1$ with the aim that $F[j]$ stores the frequency of $T[j]T[j+1]$ in $T[1..j]$. We can fill the array in σ_i scans over T_i, costing us $\mathcal{O}(n_i \sigma_i)$ time. The largest number stored in F is the most frequent bigram in T.

Heuristic 3. Finally, if the distribution of bigrams is skewed, chances are that one bigram outnumbers all others. In such a case, we can use the following algorithm to find this bigram:

Lemma 9. *Given there is a bigram in T_i ($0 \leq i \leq n$) whose frequency is higher than the sum of frequencies of all other bigrams, we can compute T_{i+1} in $\mathcal{O}(n)$ time using $\mathcal{O}(\lg n)$ bits.*

Proof. We use the Boyer–Moore majority vote algorithm [41] for finding the most frequent bigram in $\mathcal{O}(n)$ time with $\mathcal{O}(\lg n)$ bits of working space. $\square$

A practical optimization of updating F as described in Section 2.3 could be to enlarge F beyond f_k instead of keeping its size. There, after a replacement of a bigram with a non-terminal X_{i+1}, we insert those bigrams containing X_{i+1} into F whose frequencies are above t_k while discarding bigrams of the lowest frequency stored in F to keep the size of F

at f_k. Instead of discarding these bigrams, we could just let F grow. We can let F grow by using the space reserved for the frequency table F' computed in Lemma 2 (remember the definition of the constant α_i). By doing so, we might extend the lifespan of a round.

8. Conclusions

In this article, we propose an algorithm computing Re-Pair in-place in sub-quadratic time for small alphabet sizes. Our major tools are simple, which allows us to parallelize our algorithm or adapt it in the external memory model.

This paper is an extended version of our paper published in The Prague Stringology Conference 2020 [42] and our poster at the Data Compression Conference 2020 [43].

Author Contributions: Conceptualization: K.G.; formal analysis: T.I., I.F., Y.T., and D.K.; visualization: K.S.; writing: T.I. and D.K. All have authors read and agreed to the published version of the manuscript.

Funding: This work is funded by the JSPS KAKENHI Grant Numbers JP18F18120 (Dominik Köppl), 19K20213 (Tomohiro I), and 18K18111 (Yoshimasa Takabatake) and the JST CREST Grant Number JPMJCR1402 including the AIP challenge program (Keisuke Goto).

Institutional Review Board Statement: Not applicable

Informed Consent Statement: Not applicable

Data Availability Statement: Data sharing not applicable

Conflicts of Interest: The authors declare no conflict of interest.

References

1. Larsson, N.J.; Moffat, A. Offline Dictionary-Based Compression. In Proceedings of the 1999 Data Compression Conference, Snowbird, UT, USA, 29–31 March 1999; pp. 296–305.
2. Navarro, G.; Russo, L.M.S. Re-Pair Achieves High-Order Entropy. In Proceedings of the 2008 Data Compression Conference, Snowbird, UT, USA, 25–27 March 2008; p. 537
3. Kieffer, J.C.; Yang, E. Grammar-based codes: A new class of universal lossless source codes. *IEEE Trans. Inf. Theory* **2000**, *46*, 737–754.
4. Ochoa, C.; Navarro, G. RePair and All Irreducible Grammars are Upper Bounded by High-Order Empirical Entropy. *IEEE Trans. Inf. Theory* **2019**, *65*, 3160–3164.
5. Ganczorz, M. Entropy Lower Bounds for Dictionary Compression. *Proc. CPM* **2019**, *128*, 11:1–11:18.
6. Charikar, M.; Lehman, E.; Liu, D.; Panigrahy, R.; Prabhakaran, M.; Sahai, A.; Shelat, A. The smallest grammar problem. *IEEE Trans. Inf. Theory* **2005**, *51*, 2554–2576.
7. Bannai, H.; Hirayama, M.; Hucke, D.; Inenaga, S.; Jez, A.; Lohrey, M.; Reh, C.P. The smallest grammar problem revisited. *arXiv* **2019**, arXiv:1908.06428.
8. Yoshida, S.; Kida, T. Effective Variable-Length-to-Fixed-Length Coding via a Re-Pair Algorithm. In Proceedings of the 2013 Data Compression Conference, Snowbird, UT, USA, 20–22 March 2013; p. 532.
9. Lohrey, M.; Maneth, S.; Mennicke, R. XML tree structure compression using RePair. *Inf. Syst.* **2013**, *38*, 1150–1167.
10. Tabei, Y.; Saigo, H.; Yamanishi, Y.; Puglisi, S.J. Scalable Partial Least Squares Regression on Grammar-Compressed Data Matrices. In Proceedings of the SIGKDD, San Francisco, CA, USA, 13–17 August 2016; pp. 1875–1884.
11. De Luca, P.; Russiello, V.M.; Ciro Sannino, R.; Valente, L. A study for Image compression using Re-Pair algorithm. *arXiv* **2019**, arXiv 1901.10744.
12. Claude, F.; Navarro, G. Fast and Compact Web Graph Representations. *TWEB* **2010**, *4*, 16:1–16:31.
13. González, R.; Navarro, G.; Ferrada, H. Locally Compressed Suffix Arrays. *ACM J. Exp. Algorithmics* **2014**, *19*, 1.
14. Sekine, K.; Sasakawa, H.; Yoshida, S.; Kida, T. Adaptive Dictionary Sharing Method for Re-Pair Algorithm. In Proceedings of the 2014 Data Compression Conference, Snowbird, UT, USA, 26–28 March 2014; p. 425.
15. Masaki, T.; Kida, T. Online Grammar Transformation Based on Re-Pair Algorithm. In Proceedings of the 2016 Data Compression Conference (DCC), Snowbird, UT, USA, 30 March–1 April 2016; pp. 349–358.
16. Ganczorz, M.; Jez, A. Improvements on Re-Pair Grammar Compressor. In Proceedings of the 2017 Data Compression Conference (DCC), Snowbird, UT, USA, 4–7 April 2017; pp. 181–190.
17. Ziv, J.; Lempel, A. A universal algorithm for sequential data compression. *IEEE Trans. Inf. Theory* **1977**, *23*, 337–343.
18. Furuya, I.; Takagi, T.; Nakashima, Y.; Inenaga, S.; Bannai, H.; Kida, T. MR-RePair: Grammar Compression based on Maximal Repeats. In Proceedings of the 2019 Data Compression Conference (DCC), Snowbird, UT, USA, 26–29 March 2019; pp. 508–517.

19. Kärkkäinen, J.; Kempa, D.; Puglisi, S.J. Lightweight Lempel-Ziv Parsing. In Proceedings of the International Symposium on Experimental Algorithms, Rome, Italy, 5–7 June 2013; Volume 7933, pp. 139–150.
20. Goto, K. Optimal Time and Space Construction of Suffix Arrays and LCP Arrays for Integer Alphabets. In Proceedings of the Prague Stringology Conference 2019, Prague, Czech Republic, 26–28 August, 2019; pp. 111–125.
21. Manber, U.; Myers, E.W. Suffix Arrays: A New Method for On-Line String Searches. *SIAM J. Comput.* **1993**, *22*, 935–948.
22. Li, Z.; Li, J.; Huo, H. Optimal In-Place Suffix Sorting. In Proceedings of the International Symposium on String Processing and Information Retrieval, Lima, Peru, 9–11 October 2018; Volume 11147, pp. 268–284.
23. Crochemore, M.; Grossi, R.; Kärkkäinen, J.; Landau, G.M. Computing the Burrows-Wheeler transform in place and in small space. *J. Discret. Algorithms* **2015**, *32*, 44–52.
24. da Louza, F.A.; Gagie, T.; Telles, G.P. Burrows-Wheeler transform and LCP array construction in constant space. *J. Discret. Algorithms* **2017**, *42*, 14–22.
25. Kosolobov, D. Faster Lightweight Lempel-Ziv Parsing. In Proceedings of the International Symposium on Mathematical Foundations of Computer Science, Milano, Italy, 24–28 August 2015; Volume 9235, pp. 432–444.
26. Nakamura, R.; Inenaga, S.; Bannai, H.; Funamoto, T.; Takeda, M.; Shinohara, A. Linear-Time Text Compression by Longest-First Substitution. *Algorithms* **2009**, *2*, 1429–1448.
27. Bille, P.; Gørtz, I.L.; Prezza, N. Space-Efficient Re-Pair Compression. In Proceedings of the 2017 Data Compression Conference (DCC), Snowbird, UT, USA, 4–7 April 2017; pp. 171–180.
28. Bille, P.; Gørtz, I.L.; Prezza, N. Practical and Effective Re-Pair Compression. *arXiv* **2017**, arXiv:1704.08558.
29. Sakai, K.; Ohno, T.; Goto, K.; Takabatake, Y.; I, T.; Sakamoto, H. RePair in Compressed Space and Time. In Proceedings of the 2019 Data Compression Conference (DCC), Snowbird, UT, USA, 26–29 March 2019; pp. 518–527.
30. Carrascosa, R.; Coste, F.; Gallé, M.; López, G.G.I. Choosing Word Occurrences for the Smallest Grammar Problem. In Proceedings of the International Conference on Language and Automata Theory and Applications, Trier, Germany, 24–28 May 2010; Volume 6031, pp. 154–165.
31. Gage, P. A New Algorithm for Data Compression. *C Users J.* **1994**, *12*, 23–38.
32. Chan, T.M.; Munro, J.I.; Raman, V. Selection and Sorting in the "Restore" Model. *ACM Trans. Algorithms* **2018**, *14*, 11:1–11:18.
33. Williams, J.W.J. Algorithm 232—Heapsort. *Commun. ACM* **1964**, *7*, 347–348.
34. Vigna, S. Broadword Implementation of Rank/Select Queries. In Proceedings of the International Workshop on Experimental and Efficient Algorithms, Provincetown, MA, USA, 30 May–1 June 2008; Volume 5038, pp. 154–168.
35. Fredman, M.L.; Willard, D.E. Surpassing the Information Theoretic Bound with Fusion Trees. *J. Comput. Syst. Sci.* **1993**, *47*, 424–436.
36. Knuth, D.E. *The Art of Computer Programming, Volume 4, Fascicle 1: Bitwise Tricks & Techniques; Binary Decision Diagrams*, 12th ed.; Addison-Wesley: Boston, MA, USA, 2009.
37. Batcher, K.E. Sorting Networks and Their Applications. In Proceedings of the AFIPS Spring Joint Computer Conference, Atlantic City, NJ, USA, 30 April–2 May 1968; Volume 32, pp. 307–314.
38. Aggarwal, A.; Vitter, J.S. The Input/Output Complexity of Sorting and Related Problems. *Commun. ACM* **1988**, *31*, 1116–1127.
39. Jiang, S.; Larsen, K.G. A Faster External Memory Priority Queue with DecreaseKeys. In Proceedings of the 2019 Annual ACM-SIAM Symposium on Discrete Algorithms, San Diego, CA, USA, 6–9 January 2019; pp. 1331–1343.
40. Simic, S. Jensen's inequality and new entropy bounds. *Appl. Math. Lett.* **2009**, *22*, 1262–1265.
41. Boyer, R.S.; Moore, J.S. MJRTY: A Fast Majority Vote Algorithm. In *Automated Reasoning: Essays in Honor of Woody Bledsoe*; Automated Reasoning Series; Springer: Dordrecht, The Netherlands, 1991; pp. 105–118.
42. Köppl, D.; I, T.; Furuya, I.; Takabatake, Y.; Sakai, K.; Goto, K. Re-Pair in Small Space. In Proceedings of the Prague Stringology Conference 2020, Prague, Czech Republic, 31 August–2 September 2020; pp. 134–147.
43. Köppl, D.; I, T.; Furuya, I.; Takabatake, Y.; Sakai, K.; Goto, K. Re-Pair in Small Space (Poster). In Proceedings of the 2020 Data Compression Conference, Snowbird, UT, USA, 24–27 March 2020; p. 377.

Article

Subpath Queries on Compressed Graphs: A Survey

Nicola Prezza 

Department of Environmental Sciences, Informatics and Statistics, Ca' Foscari University, Dorsoduro, 3246, 30123 Venice, Italy; nicola.prezza@unive.it

Abstract: Text indexing is a classical algorithmic problem that has been studied for over four decades: given a text T, pre-process it off-line so that, later, we can quickly count and locate the occurrences of any string (the query pattern) in T in time proportional to the query's length. The earliest optimal-time solution to the problem, the suffix tree, dates back to 1973 and requires up to two orders of magnitude more space than the plain text just to be stored. In the year 2000, two breakthrough works showed that efficient queries can be achieved without this space overhead: a fast index be stored in a space proportional to the text's entropy. These contributions had an enormous impact in bioinformatics: today, virtually any DNA aligner employs compressed indexes. Recent trends considered more powerful compression schemes (dictionary compressors) and generalizations of the problem to labeled graphs: after all, texts can be viewed as labeled directed paths. In turn, since finite state automata can be considered as a particular case of labeled graphs, these findings created a bridge between the fields of compressed indexing and regular language theory, ultimately allowing to index regular languages and promising to shed new light on problems, such as regular expression matching. This survey is a gentle introduction to the main landmarks of the fascinating journey that took us from suffix trees to today's compressed indexes for labeled graphs and regular languages.

Keywords: indexing; compressed data structures; labeled graphs

---

Citation: Prezza, N. Subpath Queries on Compressed Graphs: A Survey. *Algorithms* **2021**, *14*, 14. https://doi.org/10.3390/a14010014

Received: 19 November 2020
Accepted: 3 January 2021
Published: 5 January 2021

Publisher's Note: MDPI stays neutral with regard to jurisdictional claims in published maps and institutional affiliations.

1. Introduction

Consider the classic algorithmic problem of finding the occurrences of a particular string Π (a *pattern*) in a text T. Classic algorithms, such as Karp-Rabin's [1], Boyer-Moore-Galil's [2], Apostolico-Giancarlo's [3], and Knuth-Morris-Pratt's [4], are optimal (the first only in the expected case) when both the text and the pattern are part of the query: those algorithms scan the text and find all occurrences of the pattern in linear time. What if the text is known *beforehand* and only the patterns to be found are part of the query? In this case, it is conceivable that *preprocessing* T off-line into a *fast* and *small* data structure (an *index*) might be convenient over the above on-line solutions. As it turns out, this is exactly the case. The *full-text indexing* problem (where *full* refers to the fact that we index the full set of T's substrings) has been studied for over forty years and has reached a very mature and exciting point: modern algorithmic techniques allow us to build text indexes taking a space close to that of the compressed text and able to count/locate occurrences of a pattern inside it in time proportional to the pattern's length. Note that we have emphasized two fundamental features of these data structures: query *time* and index *space*. As shown by decades of research on the topic, these two dimensions are, in fact, strictly correlated: the structure exploited to compress text is often the same that can be used also to support fast search queries on it. Recent research has taken a step forward, motivated by the increasingly complex structure of modern massive datasets: texts can be viewed as directed labeled path graphs and compressed indexing techniques can be generalized to more complex graph topologies. While compressed text indexing has already been covered in the literature in excellent books [5,6] and surveys [7–9], the generalizations of these advanced techniques to labeled graphs, dating back two decades, lack a single point of reference despite having reached a mature state-of-the-art. The goal of this survey is to introduce the (non-expert) reader to the fascinating field that studies compressed indexes for labeled graphs.

We start, in Section 2, with a quick overview of classical compressed text indexing techniques: compressed suffix arrays and indexes for repetitive collections. This section serves as a self-contained warm-up to fully understand the concepts contained in the next sections. Section 3 starts with an overview of the problem of compressing graphs, with a discussion on known lower bounds to the graph indexing problem, and with preliminary solutions based on hypertext indexing. We then introduce prefix sorting and discuss its extensions to increasingly complex graph topologies. Section 3.5 is devoted to the problem of indexing trees, the first natural generalization of classical labeled paths (strings). Most of the discussion in this section is spent on the eXtended Burrows-Wheeler Transform (XBWT), a tree transformation reflecting the co-lexicographic order of the root-to-node paths on the tree. We discuss how this transformation naturally supports subpath queries and compression by generalizing the ideas of Section 2. Section 3.6 adds further diversity to the set of indexable graph topologies by discussing the cases of sets of disjoints cycles and de Bruijn graphs. All these cases are finally generalized in Section 3.7 with the introduction of *Wheeler graphs*, a notion capturing the idea of totally-sortable labeled graph. This section discusses the problem of compressing and indexing Wheeler graphs, as well as recognizing and sorting them. We also spend a paragraph on the fascinating bridge that this technique builds between compressed indexing and regular language theory by briefly discussing the elegant properties of *Wheeler languages*: regular languages recognized by finite automata in which state transition is a Wheeler graph. In fact, we argue that a useful variant of graph indexing is *regular language indexing*: in many applications (for example, computational pan-genomics [10]) one is interested in indexing the set of strings read on the paths of a labeled graphs, rather than indexing a fixed graph topology. As a matter of fact, we will see that the complexity of those two problems is quite different. Finally, Section 3.8 further generalizes prefix-sorting to any labeled graph, allowing us to index any regular language. The key idea of this generalization is to abandon total orders (of prefix-sorted states) in favor of partial ones. We conclude our survey, in Section 4, with a list of open challenges in the field.

Terminology

A string S of length n over alphabet Σ is a sequence of n elements from Σ. We use the notation $S[i]$ to indicate the i-th element of S, for $1 \le i \le n$. Let $a \in \Sigma$ and $S \in \Sigma^*$. We write Sa to indicate the concatenation of S and a. A string can be interpreted as an edge-labeled graph (a path) with $n + 1$ nodes connected by n labeled edges. In general, a *labeled graph* $G = (V, E, \Sigma, \lambda)$ is a directed graph with set of nodes V, set of edges $E \subseteq V \times V$, alphabet Σ, and labeling function $\lambda : E \to \Sigma$. Quantities $n = |V|$ and $e = |E|$ will indicate the number of nodes and edges, respectively. We assume to work with an *effective* alphabet Σ of size σ, that is, every $c \in \Sigma$ is the label of some edge. In particular, $\sigma \le e$. We moreover assume the alphabet to be totally ordered by an order we denote with $\le$, and write $a < b$ when $a \le b$ and $a \ne b$. In this survey, we consider two extensions of $\le$ to strings (and, later, to labeled graphs). The *lexicographic order* of two strings aS and $a'S'$ is defined as $a'S' < aS$ if and only if either (i) $a' < a$ or (ii) $a = a'$ and $S' < S$ hold. The empty string ϵ is always smaller than any non-empty string. Symmetrically, the *co-lexicographic order* of two strings Sa and $S'a'$ is defined as $S'a' < Sa$ if and only if either (i) $a' < a$ or (ii) $a = a'$ and $S' < S$ hold.

This paper deals with the indexed pattern matching problem: preprocess a text so that, later all text occurrences of any query pattern $\Pi \in \Sigma^m$ of length m can be efficiently counted and located. These queries can be generalized to labeled graphs; we postpone the exact definition of this generalization to Section 3.

2. The Labeled Path Case: Indexing Compressed Text

Consider the text reported in Figure 1. For reasons that will be made clear later, we append a special symbol $ at the end of the text, and assume that $ is lexicographically smaller than all other characters. Note that we assign a different color to each distinct letter.

$$\mathcal{T} \;=\; \begin{matrix} A & T & A & T & A & G & A & T & \$ \\ 1 & 2 & 3 & 4 & 5 & 6 & 7 & 8 & 9 \end{matrix}$$

Figure 1. Running example used in this section.

The question is: how can we design a data structure that permits to efficiently find the exact occurrences of (short) strings inside $\mathcal{T}$? In particular, one may wish to *count* the occurrences of a pattern (for example, $count(\mathcal{T}, \text{"}AT\text{"}) = 3$) or to *locate* them (for example, $locate(\mathcal{T}, \text{"}AT\text{"}) = 1, 3, 7$). As we will see, there are two conceptually different ways to solve this problem. The first, sketched in Section 2.1, is to realize that every occurrence of $\Pi = \text{"}AT\text{"}$ in $\mathcal{T}$ is a prefix of a text suffix (for example: occurrence at position 7 of "AT" is a prefix of the text suffix "AT"). The second, sketched in Section 2.2, is to partition the text into non-overlapping *phrases* appearing elsewhere in the text and divide the problem into the two sub-problems of (i) finding occurrences that overlap two adjacent phrases and (ii) finding occurrences entirely contained in a phrase. Albeit conceptually different, both approaches ultimately resort to *suffix sorting*: sorting lexicographically a subset of the text's suffixes. The curious reader can refer to the excellent reviews of Mäkinen and Navarro [7] and Navarro [8,9] for a much deeper coverage of the state-of-the-art relative to both approaches.

2.1. The Entropy Model: Compressed Suffix Arrays

The sentence *every occurrence of Π is the prefix of a suffix of $\mathcal{T}$* leads very quickly to a simple and time-efficient solution to the full-text indexing problem. Note that a suffix can be identified by a text position: for example, suffix "GAT$" corresponds to position 6. Let us sort text suffixes in *lexicographic order*. We call the resulting integer array the *suffix array* (SA) of $\mathcal{T}$; see Figure 2. This time, we color *text positions i* according to the color of letter $\mathcal{T}[i]$. Note that colors (letters) get clustered, since suffixes are sorted lexicographically.

$$SA \;=\; \begin{matrix} 9 & 5 & 7 & 3 & 1 & 6 & 8 & 4 & 2 \\ \$ & A & A & A & A & G & T & T & T \\ & G & T & T & T & A & \$ & A & A \\ & A & \$ & A & A & T & & G & T \\ & T & & G & T & \$ & & A & A \\ & \$ & & A & A & & & T & G \\ & & & T & G & & & \$ & A \\ & & & \$ & A & & & & T \\ & & & & T & & & & \$ \\ & & & & \$ & & & & \end{matrix}$$

Figure 2. Suffix Array (SA) of the text $\mathcal{T}$ of Figure 1. Note: we store only array SA and the text $\mathcal{T}$, not the actual text suffixes.

The reason for appending $ at the end of the text is that, in this way, no suffix prefixes another suffix. Suffix arrays were independently discovered by Udi Manber and Gene Myers in 1990 [11] and (under the name of PAT array) by Gonnet, Baeza-Yates, and Snider in 1992 [12,13]. An earlier solution, the *suffix tree* [14], dates back to 1973 and is more time-efficient, albeit much less space efficient (by a large constant factor). The idea behind the suffix tree is to build the trie of all text's suffixes, replacing unary paths with pairs of pointers to the text.

The suffix array SA, when used together with the text $\mathcal{T}$, is a full-text index: in order to count/locate occurrences of a pattern Π, it is sufficient to binary search SA, extracting characters from $\mathcal{T}$ to compare Π with the corresponding suffixes during search. This solution permits to count the *occ* occurrences of a pattern Π of length m in a text $\mathcal{T}$ of length n in $O(m \log n)$ time, as well as to report them in additional optimal $O(occ)$ time. Letting the alphabet size being denoted by σ, our index requires $n \log \sigma + n \log n$ bits to be stored (the first component for the text and the second for the suffix array).

Note that the suffix array cannot be considered a *small* data structure. On a constant-sized alphabet, the term $n \log n$ is asymptotically larger than the text. The need for processing larger and larger texts made this issue relevant by the end of the millennium. As an instructive example, consider indexing the Human genome ($\approx$3 billion DNA bases). DNA sequences can be modeled as strings over an alphabet of size four ($\{A, C, G, T\}$); therefore, the Human genome takes less than 750 MiB of space to be stored using two bits per letter. The suffix array of the Human genome requires, on the other hand, about 11 GiB.

Although the study of *compressed text indexes* had begun before the year 2000 [15], the turn of the millennium represented a crucial turning point for the field: two independent works by Ferragina and Manzini [16] and Grossi and Vitter [17] showed how to *compress* the suffix array.

Consider the integer array ψ of length n defined as follows. Given a suffix array position i containing value (text position) $j = SA[i]$, the cell $\psi[i]$ contains the suffix array position i' containing text position $(j \mod n) + 1$ (we treat the string as circular). More formally: $\psi[i] = SA^{-1}[(SA[i] \mod n) + 1]$. See Figure 3.

$$
\begin{array}{lccccccccc}
SA & = & 9 & 5 & 7 & 3 & 1 & 6 & 8 & 4 & 2 \\
\psi & = & 5 & 6 & 7 & 8 & 9 & 3 & 1 & 2 & 4 \\
 & & 1 & 2 & 3 & 4 & 5 & 6 & 7 & 8 & 9
\end{array}
$$

Figure 3. Array ψ.

Note the following interesting property: equally-colored values in ψ are increasing. More precisely: ψ-values corresponding to suffixes beginning with the same letter form increasing subsequences. To understand why this happens, observe that $\psi[i]$ takes us from a suffix $\mathcal{T}[SA[i], \dots, n]$ to suffix $\mathcal{T}[SA[i] + 1, \dots, n]$ (let us exclude the case $SA[i] = n$ in order to simplify our formulas). Now, assume that $i < j$ and $\mathcal{T}[SA[i], \dots, n]$ and $\mathcal{T}[SA[j], \dots, n]$ begin with the same letter (that is: i and j have the same color in Figure 3). Since $i < j$, we have that $\mathcal{T}[SA[i], \dots, n] < \mathcal{T}[SA[j], \dots, n]$ (lexicographically). But then, since the two suffixes begin with the same letter, we also have that $\mathcal{T}[SA[i] + 1, \dots, n] < \mathcal{T}[SA[j] + 1, \dots, n]$, i.e., $\psi[i] < \psi[j]$.

Let us now store just the differences between consecutive values in each increasing subsequence of ψ. We denote this new array of differences as $\Delta(\psi)$. See Figure 4 for an example.

$$
\begin{array}{lccccccccc}
SA & = & 9 & 5 & 7 & 3 & 1 & 6 & 8 & 4 & 2 \\
\Delta(\psi) & = & 5 & 6 & 1 & 1 & 1 & 3 & 1 & 1 & 2 \\
 & & 1 & 2 & 3 & 4 & 5 & 6 & 7 & 8 & 9
\end{array}
$$

Figure 4. The differential array $\Delta(\psi)$.

Two remarkable properties of $\Delta(\psi)$ are that, using an opportune encoding for its elements, this sequence:

1. supports accessing any $\psi[i]$ in constant time [17], and
2. can be stored in $nH_0 + O(n)$ bits of space, where $H_0 = \sum_{c \in \Sigma}(n_c/n) \log(n/n_c)$ is the zero-order empirical entropy of $\mathcal{T}$ and n_c denotes the number of occurrences of $c \in \Sigma$ in $\mathcal{T}$.

The fact that this strategy achieves compression is actually not hard to prove. Consider any integer encoding (for example, Elias' delta or gamma) capable to represent any integer x in $O(1 + \log x)$ bits. By the concavity of the logarithm function, the inequality $\sum_{i=1}^{m} \log(x_i) \leq m \cdot \log\left(\frac{\sum_{i=1}^{m} x_i}{m}\right)$ holds for any integer sequence $x_1, \dots, x_m$. Now, note that the sub-sequence $x_1, \dots, x_{n_c}$ of $\Delta(\psi)$ corresponding to letter $c \in \Sigma$ has two properties:

it has exactly n_c terms and $\sum_{i=1}^{n_c} x_i \leq n$. It follows that the encoded sub-sequence takes (asymptotically) $\sum_{i=1}^{n_c}(1 + \log(x_i)) \leq n_c \cdot \log(n/n_c) + n_c$ bits. By summing this quantity over all the alphabet's characters, we obtain precisely $O(n(H_0 + 1))$ bits. Other encodings (in particular, Elias-Fano dictionaries [18,19]) can achieve the claimed $nH_0 + O(n)$ bits while supporting constant-time random access.

The final step is to recognize that ψ moves us forward (by one position at a time) in the text. This allows us to extract suffixes *without using the text*. To achieve this, it is sufficient to store in one array $F = \$AAAAGTTT$ (using our running example) the first character of each suffix. This can be done in $O(n)$ bits of space (using a bitvector) since those characters are sorted and we assume the alphabet to be effective. The i-th text suffix (in lexicographic order) is then $F[i], F[\psi[i]], F[\psi^{(2)}[i]], F[\psi^{(3)}[i]], \ldots$, where $\psi^{(\ell)}$ indicates function ψ applied ℓ times to its argument. See Figure 5. Extracting suffixes makes it possible to implement the binary search algorithm discussed in the previous section: the *Compressed Suffix Array* (CSA) takes compressed space and enables us finding all pattern's occurrences in a time proportional to the pattern length. By adding a small sampling of the suffix array, one can use the same solution to compute any value $SA[i]$ in polylogarithmic time without asymptotically affecting the space usage.

		1	2	3	4	5	6	7	8	9
ψ	$=$	5	6	7	8	9	3	1	2	4
F	$=$	$\underline{\$}$	$\underline{A}$	$\underline{A}$	$\underline{A}$	$\underline{A}$	$\underline{G}$	$\underline{T}$	$\underline{T}$	$\underline{T}$
			G	T	T	T	A	$	A	A
			A	$	A	A	T		G	T
			T		G	T	$		A	A
			$		A	A			T	G
					T	G			$	A
					$	A				T
						T				$
						$				

Figure 5. Compressed Suffix Array (CSA): we store the delta-encoded ψ and the first letter (underlined) of each suffix (array F).

Another (symmetric) philosophy is to exploit the inverse of function ψ. These indexes are based on the *Burrows-Wheeler transform* (BWT) [20] and achieve high-order compression [16]. We do not discuss BWT-based indexes here since their generalizations to labeled graphs will be covered in Section 3. For more details on entropy-compressed text indexes, we redirect the curious reader to the excellent survey of Mäkinen and Navarro [7].

2.2. The Repetitive Model

Since their introduction in the year 2000, entropy-compressed text indexes have had a dramatic impact in domains, such as bioinformatics; the most widely used DNA aligners, Bowtie [21] and BWA [22], are based on compressed indexes and can align thousands of short DNA fragments per second on large genomes while using only compressed space in RAM during execution. As seen in the previous section, these indexes operate within a space bounded by the text's empirical entropy [16,17]. Entropy, however, is insensitive to long repetitions: the entropy-compressed version of $\mathcal{T} \cdot \mathcal{T}$ (that is, text $\mathcal{T}$ concatenated with itself) takes at least twice the space of the entropy-compressed $\mathcal{T}$ [23]. There is a simple reason for this fact: by its very definition (see Section 2.1), the quantity H_0 depends only on the characters' relative frequencies in the text. Since characters in $\mathcal{T}$ and $\mathcal{T} \cdot \mathcal{T}$ have the same relative frequencies, it follows that their entropies are the same. The claim follows, being the length of $\mathcal{T} \cdot \mathcal{T}$ twice the length of $\mathcal{T}$.

While the above reasoning might seem artificial, the same problem arises on texts composed of large repetitions. Today, most large data sources, such as DNA sequencers and the web, follow this new model of *highly repetitive data*. As a consequence, entropy-compressed text indexes are no longer able to keep pace with the exponentially-increasing rate at which this data is produced, as very often the mere index size exceeds the RAM limit. For this reason, in recent years more powerful compressed indexes have emerged; these are based on the Lempel-Ziv factorization [23], the run-length Burrows-Wheeler Transform (BWT) [24–26], context-free grammars [27], string attractors [28,29] (combinatorial objects generalizing those compressors), and more abstract measures of repetitiveness [30]. The core idea behind these indexes (with the exception of the run-length BWT; see Section 3) is to partition the text into phrases that occur also elsewhere and to use *geometric* data structures to locate pattern occurrences.

To get a more detailed intuition of how these indexes work, consider the Lempel-Ziv'78 (LZ78) compression scheme. The LZ78 factorization breaks the text into phrases with the property that each phrase extends by one character a previous phrase; see the example in Figure 6 (top). The mechanism we are going to describe works for any factorization-based compressor (also called *dictionary compressors*). A factorization-based index keeps a two-dimensional geometric data structure storing a labeled point (x, y, ℓ) for each phrase y, where x is the reversed text prefix preceding the phrase and ℓ is the first text position of the phrase. Efficient techniques not discussed here (in particular, mapping to *rank space*) exist to opportunely reduce x and y to small integers preserving the lexicographic order of the corresponding strings (see, for example, Kreft and Navarro [23]). For example, such a point in Figure 6 is $(GCA, CG, 4)$, corresponding to the phrase break between positions 3 and 4. To understand how *locate* queries are answered, consider the pattern $\Pi = CAC$. All occurrences of Π crossing a phrase boundary can be split into a prefix and a suffix, aligned on the phrase boundary. Let us consider the split $CA|C$ (all possible splits have to be considered for the following algorithm to be complete). If Π occurs in $\mathcal{T}$ with this split, then C will be a prefix of some phrase y, and CA will be a suffix of the text prefix preceding y. We can find all phrases y with this property by issuing a four-sided range query on our grid as shown in Figure 6 (bottom left). This procedure works since C and AC (that is, CA reversed) define ranges on the horizontal and vertical axes: these ranges contain all phrases prefixed by C and all reversed prefixes prefixed by AC (equivalently, all prefixes suffixed by CA), respectively. In Figure 6 (bottom left), the queried rectangle contains text positions 13 and 11. By subtracting from those numbers the length of the pattern's prefix (in this example, $2 = |CA|$), we discover that $\Pi = CAC$ occurs at positions 11 and 9 crossing a phrase with split $CA|C$.

With a similar idea (that is, resorting again to geometric data structures), one can recursively track occurrences completely contained inside a phrase; see the caption of Figure 6 (bottom right). Let b be the number of phrases in the parse. Note that our geometric structures store overall $O(b)$ points. On repetitive texts, the smallest possible number b of phrases of such a parse can be *constant*. In practice, on repetitive texts b (or any of its popular approximations [31]) is orders of magnitude smaller than the entropy-compressed text [23,24]. The fastest existing factorization-based indexes to date can locate all *occ* occurrences of a pattern $\Pi \in \Sigma^m$ in $O(b \text{ polylog } n)$ bits of space and *optimal* $O(m + occ)$ time [32]. For more details on indexes for repetitive text collections, the curious reader can refer to the excellent recent survey of Navarro [8,9].

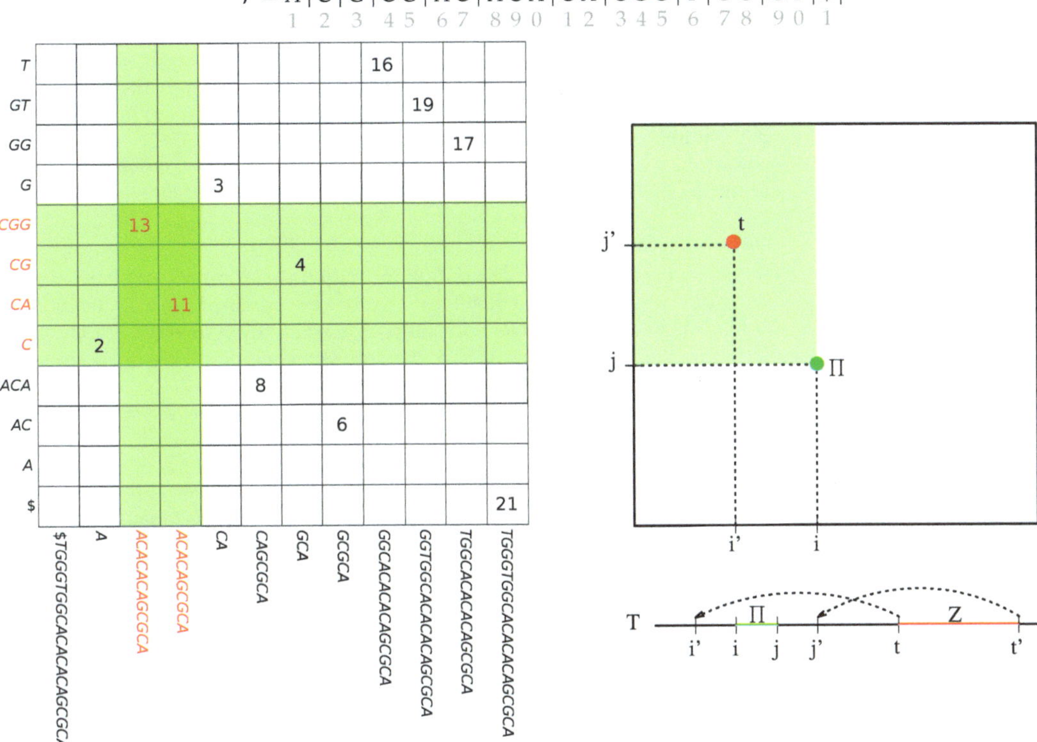

Figure 6. (Top) Parsed text according to the LZ78 factorization. Text indices are shown in gray. **(Bottom left)** four-sided geometric data structure storing one labeled point (x, y, ℓ) per phrase y, where x is the reversed text prefix preceding phrase y and ℓ is the first position of the phrase. **(Bottom right)** given a pattern occurrence $\Pi = \mathcal{T}[i, j]$ (green dot), we can locate all phrases that completely copy it. For each phrase $\mathcal{T}[t, t + \ell - 1]$ for which source is $\mathcal{T}[i', i' + \ell - 1]$, a labeled point $(i', i' + \ell - 1, t)$ is inserted in the data structure. In the example: phrase $Z = \mathcal{T}[t, t']$, red dot, copies $\mathcal{T}[i', j']$ which completely contains Π. Note that $\Pi = \mathcal{T}[i, j]$ defines the query, while each phrase generates a point that is stored permanently in the geometric structure.

3. Indexing Labeled Graphs and Regular Languages

Interestingly enough, most techniques seen in the previous section extend to more structured data. In particular, in this section, we will work with *labeled graphs*, and discuss extensions of these results to *regular languages* by interpreting finite automata as labeled graphs. We recall that a labeled graph is a quadruple (V, E, Σ, λ) where V is a set of $n = |V|$ nodes, $E \subseteq V \times V$ is a set of $e = |E|$ directed edges, Σ is the alphabet and $\lambda : E \to \Sigma$ is a labeling function assigning a label to each edge. Let $P = (u_{i_1}, u_{i_2}), (u_{i_2}, u_{i_3}), \ldots, (u_{i_k}, u_{i_{k+1}})$ be a path of length k. We extend function λ to paths as $\lambda(P) = \lambda((u_{i_1}, u_{i_2})) \cdot \lambda((u_{i_2}, u_{i_3})) \cdots \lambda((u_{i_k}, u_{i_{k+1}}))$, and say that $\lambda(P)$ is the string labeling path P. A node u is *reached by a path labeled* $\Pi \in \Sigma^*$ if there exists a path $P = (u_{i_1}, u_{i_2}), (u_{i_2}, u_{i_3}), \ldots, (u_{i_k}, u)$ ending in u such that $\lambda(P) = \Pi$.

The starting point is to observe that texts are nothing but labeled *path graphs*. As it turns out, there is nothing special about paths that we cannot generalize to more complex topologies. We adopt the following natural generalizations of count and locate queries:

- *Count*: given a pattern (a string) $\Pi \in \Sigma^m$, return the number of nodes reached by a path labeled with Π.

- *Locate*: given a pattern Π, return a representation of all nodes reached by a path labeled with Π.

We will refer to the set of these two queries with the term *subpath queries*. The node representation returned by locate queries could be an arbitrary numbering. This solution, however, has the disadvantage of requiring at least $n \log n$ bits of space just for the labels, n being the number of nodes. In the following subsections, we will discuss more space-efficient solutions based on the idea of returning a more regular labeling (e.g., the DFS order of the nodes).

3.1. Graph Compression

We start with a discussion of existing techniques for compressing labeled graphs. Lossless graph compression is a vast topic that has been treated more in detail in other surveys [33,34]. Since here we deal with subpath queries on compressed graphs, we only discuss compression techniques that have been shown to support these queries on special cases of graphs (mainly paths and trees).

Note that, differently from the string domain, the goal is now to compress two components: the labels and the graph's topology. Labels can be compressed by extending the familiar entropy model to graphs. The idea here is to simply count the frequencies of each character in the multi-set of all the edges' labels. By using an encoding, such as Huffman's, the zero-order empirical entropy of the labels can be approached. One can take a step further and compress labels to their high-order entropy. As noted by Ferragina et al. [35], if the topology is a tree then we can use a different zero-order model for each context of length k (that is, for each distinct labeled path of length k) preceding a given edge. This is the same observation used to compress texts: it is much easier to predict a particular character if we know the k letters (or the path of length k) that precede it. The entropy model is already very effective in compressing the labels component. The most prominent example of entropy-compressed tree index is the eXtended Burrows Wheeler transform [35], covered more in detail in Section 3.5. As far as the topology is concerned, things get more complicated. The topology of a tree with n nodes can be represented in $2n$ bits via its balanced-parentheses representation. Ferres et al. [36] used this representation and the idea that planar graphs are fully specified by a spanning tree of the graph and one of its dual to represent planar graphs in just $4n$ bits per node. Recently, Chakraborty et al. proposed succinct representations for finite automata [37]. General graph topologies are compressed well, but without worst-case guarantees, by techniques, such as K^2 trees [38]. Further compression can be achieved by extending the notion of entropy to the graph's topology, as shown by Jansson et al. [39], Hucke et al. [40], and Ganczorz [41] for the particular case of trees. As it happens with text entropy, their measures work well under the assumption that the tree topology is not extremely repetitive. See Hucke et al. [42] for a systematic comparison of tree entropy measures. In the repetitive case, more advanced notions of compression need to be considered. Straight-line grammars [43,44], that is, context-free grammars generating a single graph, are the gold-standard for capturing such repetitions. These representations can be augmented to support constant-time traversal of the compressed graph [45] but do not support indexing queries on topologies being more complex than paths. Other techniques (defined only for trees) include the Lempel-Ziv factorization of trees [46] and top trees [47]. Despite being well-suited for compressing repetitive tree topologies, these representations also are not (yet) able to support efficient indexing queries on topologies more complex than paths so we do not discuss them further. The recent *tunneling* [48] and *run-length XBW Transform* [49] compression schemes are the first compressed representations for repetitive topologies supporting count queries (the latter technique supports also locate queries, but it requires additional linear space). These techniques can be applied to Wheeler graphs [50], as well, and are covered more in detail in Sections 3.5 and 3.7.

3.2. Conditional Lower Bounds

Before diving into compressed indexes for labeled graphs, we spend a paragraph on the complexity of matching strings on labeled graphs. The main results in this direction are due to Backurs and Indyk [51] and to Equi et al. [52–54], and we considered the on-line version of the problem: both pre-processing and query time are counted towards the total running time. The former work [51] proved that, unless the Strong Exponential Time Hypothesis (SETH) [55] is false, *in the worst case* no algorithm can match a regular expression of size e against a string of length m in time $O((m \cdot e)^{1-\delta})$, for any constant $\delta > 0$. Since regular expressions can be converted into NFAs of the same asymptotic size, this result implies a quadratic conditional lower-bound to the graph pattern matching problem. Equi et al. [54] improved this result to include graphs of maximum node degree equal to two and deterministic directed acyclic graphs. Another recent paper from Potechin and Shallit [56] established a similar hardness result for the problem of determining whether a NFA accepts a given word: they showed that, provided that the NFA is sparse, a sub-quadratic algorithm for the problem would falsify SETH. All these results are based on reductions from the orthogonal vectors problem: find two orthogonal vectors in two given sets of binary d-dimensional vectors (one can easily see the similarity between this problem and string matching). The orthogonal vectors hypothesis (OV) [57] states that the orthogonal vectors problem cannot be solved in strongly subquadratic time. A further improvement has recently been made by Gibney [58], who proved that even shaving logarithmic factors from the running time $O(m \cdot e)$ would yield surprising new results in complexity theory, as well as falsifying an hypothesis recently made by Abboud and Bringmann in Reference [59] on the fine-grained complexity of SAT solvers. As noted above, in the context of our survey, these lower bounds imply that the sum between the construction and query times for a graph index cannot be sub-quadratic unless important conjectures in complexity theory fail. These lower bounds, however, do not rule out the possibility that a graph index could support efficient (say, subquadratic) queries at the cost of an expensive (quadratic or more) index construction algorithm. The more recent works of Equi et al. [52,53] addressed precisely this issue: they proved that no index that can be built in polynomial time ($O(e^\alpha)$ for any constant $\alpha \geq 1$) can guarantee strongly sub-quadratic query times, that is, $O(e^\delta m^\beta)$ query time for any constants $\delta < 1$ or $\beta < 1$. This essentially settles the complexity of the problem: assuming a reasonably fast (polynomial time) index construction algorithm, subpath-query times need to be at least quadratic ($O(m \cdot e)$) in the worst case. Since these bounds are matched by existing on-line algorithms [60], one may wonder what is the point of studying graph indexes at all. As we will see, the answer lies in parameterized complexity: even though pattern matching on graphs/regular expressions is hard *in the worst case*, it is indeed possible to solve the problem efficiently in particular cases (for example, trees [35] and particular regular expressions [51]). Ultimately, in Section 3.8 we will introduce a complete hierarchy of labeled graphs capturing the hardness of the problem.

3.3. Hypertext Indexing

Prior to the predominant prefix-sorting approach that we are going to discuss in detail in the next subsections, the problem of solving indexed path queries on labeled graphs has been tackled in the literature by resorting to geometric data structures [61,62]. These solutions work in the *hypertext model*: the objects being indexed are *node-labeled* graphs $G = (V, E, \Sigma, \lambda)$, where function $\lambda : V \rightarrow \Sigma^*$ associates a string to each node (note the difference with our edge-labeled model, where each *edge* is labeled with a *single* character). Let $P = (u_{i_1}, u_{i_2}), (u_{i_2}, u_{i_3}), \ldots, (u_{i_{k-1}}, u_{i_k})$ be a path in the graph spanning k (not necessarily distinct) nodes. With $\lambda(P)$ we denote the string $\lambda(P) = \lambda(u_{i_1}) \cdot \lambda(u_{i_2}) \cdots \lambda(u_{i_k})$. In the hypertext indexing problem, the goal is to build an index over G able to quickly support locate queries on the paths of G: given a pattern Π, determine the positions in the graph (node and starting position in the node) where an occurrence of Π starts. This labeled graph model is well suited for applications where the strings labeling each node are very long (for example, a transcriptome), in which case the label component (rather than the graph's topology) dominates the data structure's space. Both solutions discussed in Reference [61,62] resort to geometric data structures. First, a classic text index (for example, a compressed suffix array) is built over the concatenation $\lambda(u_1) \cdot \# \cdots \# \cdot \lambda(u_n)$ of the strings labeling all the graph's nodes $u_1, \ldots, u_n$. The labels are separated by a symbol # not appearing elsewhere in order to prevent false matches. Pattern occurrences completely contained inside some $\lambda(u_i)$ are found using this compressed index. Then, a geometric structure is used to "glue" nodes connected by an edge: for each edge (u, v), a two-dimensional point $(\overleftarrow{\lambda(u)}, \lambda(v))$ is added to a grid as the one shown in Figure 6 (bottom left), where $\overleftarrow{w}$ denotes the reverse of string w. Again, rank space techniques are used to reduce the points $(\overleftarrow{\lambda(u)}, \lambda(v))$ to integer pairs. Then, pattern occurrences spanning *one* edge are found by issuing a four-sided geometric query for each possible pattern split as seen in Section 2.2. The main issue with these solutions is that they cannot *efficiently* locate pattern occurrences spanning two or more edges. A tweak consists in using a seed-and-extend strategy: the pattern Π is split in short fragments of some length t, each of which is searched separately in the index. If $\lambda(u_i) \geq t$ for each node u_i, then this solution allows locating all pattern occurrences (including those spanning two or more edges). This solution, however, requires to visit the whole graph in the worst case. On realistic DNA datasets this problem is mitigated by the fact that, for large enough t, the pattern's fragments of length t are "good anchors" and are likely to occur only within occurrences of Π.

As seen, hypertext indexes work well under the assumption that the strings labeling the nodes are very long. However, this assumption is often not realistic: for instance, in pan-genomics applications a single-point mutation in the genome of an individual (that is, a substitution or a small insertion/deletion) may introduce very short edges deviating from the population's reference genome. Note that here we have only discussed *indexed* solutions. The on-line case (pattern matching on hypertext without pre-processing) has been thoroughly studied in several other works; see Reference [60,63,64].

3.4. Prefix Sorting: Model and Terminology

A very natural approach to solve efficiently the graph indexing problem on arbitrary labeled graphs is to generalize suffix sorting (Actually, for technical reasons we will use the symmetric *prefix sorting*) from strings to labeled graphs. In the following, we will make a slight simplification and work with topologies corresponding to the transition function of nondeterministic finite automata (NFAs). In particular, we will assume that:

1. there is only one node, deemed the *source state* (or start state), without incoming edges, and
2. any state is reachable from the source state.

We will use the terms *node* and *state* interchangeably. For reasons that will become clear later, we will moreover assume that the set of characters labeling the incoming edges of any node is a singleton. These assumptions are not too restrictive. It is not hard to see that these NFAs can, in fact, recognize any regular language [65] (in particular, any NFA can be transformed in polynomial time to meet these requirements).

We make a little twist and replace the lexicographic order of suffixes with the symmetric co-lexicographic order of prefixes. This turns out to be much more natural when dealing with NFAs, as we will sort states according to the co-lexicographic order of the strings labeling paths connecting the source with each state. As additional benefits of these choices:

1. we will be able to search strings *forward* (that is, left-to-right). In particular, this will enable testing membership of words in the language recognized by an indexed NFA in a natural on-line (character-by-character) fashion (traditionally, BWT-based text indexes are based on suffix sorting and support *backward* search of patterns in the text).

2. We will be able to transfer powerful language-theoretic results to the field of compressed indexing. For example, a co-lexicographic variant of the Myhill-Nerode theorem [66] will allow us to minimize the number of states of indexable DFAs.

3.5. Indexing Labeled Trees

Kosaraju [67] was the first to extend prefix sorting to a very simple class of labeled graphs: labeled trees. To do so, he defined the suffix tree of a (reversed) trie. To understand this idea, we are now going to describe a simplified data structure: the prefix array of a labeled tree.

3.5.1. The Prefix Array of a Labeled Tree

Consider the list containing the tree's nodes sorted by the co-lexicographic order of the paths connecting them to the root. We call this array the *prefix array* (PA) of the tree. See Figure 7 for an example: node 5 comes before node 8 in the ordering because the path connecting the root to 5, "aac", is co-lexicographically smaller than the path connecting the root to 8, "bc". This node ordering is uniquely defined if the tree is a trie (that is, if each node has at most one outgoing edge per label). Otherwise, the order of nodes reached by the same root-to-node path can be chosen arbitrarily.

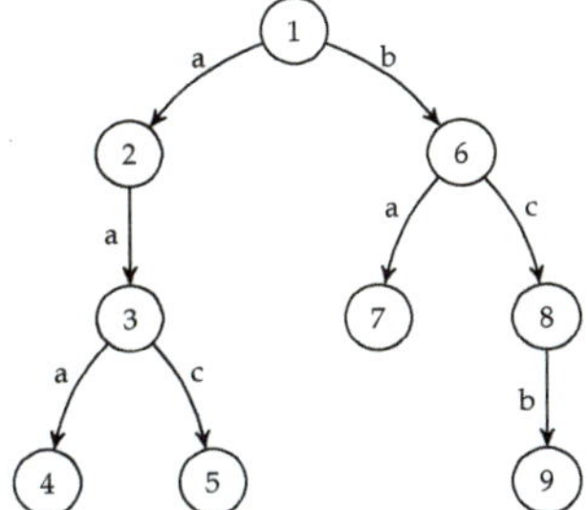

Figure 7. Example of labeled tree. Nodes have been enumerated in pre-order. The list of nodes sorted by the co-lexicographic order of the paths connecting them to the root (that is, the prefix array (PA) of the tree) is: 1, 2, 3, 4, 7, 6, 9, 5, 8.

A nice property of the prefix array of a tree is that, together with the tree itself, it is already an index; in fact, it is a straightforward generalization of the suffix array SA introduced in Section 2 (in this case, we call it "prefix array" since we are sorting prefixes of root-to-leaves paths). Since nodes are sorted co-lexicographically, the list of nodes reached by a path labeled with a given pattern Π can be found by binary search. At each search step, we jump on the corresponding node on the tree and compare Π with the labels extracted on the path connecting the node with the root. This procedure allows counting all nodes reached by Π, as well as subsequently reporting them in optimal constant time each, in $O(|\Pi| \log n)$ time, n being the number of nodes.

3.5.2. The XBW Transform

One disadvantage of the prefix array is that, like the SA seen in Section 2.1, it does not achieve compression. In addition to the $n \log n$ bits for the prefix array itself, the above binary-search strategy also requires us to navigate the tree (for which topology and labels must therefore be kept in memory). This is much more space than the labels, which require $n \log \sigma$ bits when stored in plain format, and the tree topology, which can be stored in just $2n$ bits (for example, in balanced-parentheses sequence representation). In 2005, Ferragina et al. [35,68] observed that this space overhead is not necessary: an efficient search machinery can be fit into a space proportional to the entropy-compressed edge labels, plus the succinct $(2n + o(n)$ bits) tree's topology. Their structure takes the name *XBW tree transform* (XBWT in the following), and is a compressed tree representation natively supporting subpath search queries. Consider the co-lexicographic node ordering of Figure 7, denote with $child(u)$ the multi-set of outgoing labels of node u (for the leaves, $child(u) = \emptyset$), and with $\lambda(u)$ the incoming label of node u (for the root, $\lambda(u) = \$$, where as usual $\$$ is lexicographically smaller than all other characters). Let $u_1, \ldots, u_n$ be the prefix array of the tree. The XBWT is the pair of sequences $IN = \lambda(u_1) \ldots \lambda(u_n)$ and $OUT = child(u_1), \ldots, child(u_n)$. See Figure 8 for an example based on the tree of Figure 7. For simplicity, the tree of this example is a trie (that is, a deterministic tree). All of the following reasoning, however, applies immediately to arbitrary labeled trees.

i	1	2	3	4	5	6	7	8	9
PA	1	2	3	4	7	6	9	5	8
IN	$	a	a	a	a	b	b	c	c
OUT	a	a	a			a			
	b								b
			c			c			

Figure 8. Prefix array (PA) and sequences IN and OUT forming the eXtended Burrows-Wheeler Transform (XBWT). IN (incoming labels) is the sequence of characters labeling the incoming edge of each node, while OUT (outgoing labels) is the sequence of multi-sets containing the characters labeling the outgoing edges of each node.

3.5.3. Inverting the XBWT

As it turns out, the tree can be reconstructed from just OUT. To prove this, we show that the XBWT representation can be used to perform a tree visit. First, we introduce the key property at the core of the XBWT: edges' labels appear in the same order in IN and OUT, that is, the i-th occurrence (counting from left to right) of character $c \in \Sigma$ in IN corresponds to the same edge of the i-th occurrence c in OUT (the order of characters inside each multi-set of OUT is not relevant for the following reasoning to work). For example, consider the fourth 'a' in IN, appearing at $IN[5]$. This label corresponds to the incoming edge of node 7, that is, to edge $(6,7)$. The fourth occurrence of 'a' in OUT appears in $OUT[6]$, corresponding to the outgoing edges of node 6. By following the edge labeled 'a' from node 6, we reach exactly node 7, that is, this occurrence of 'a' labels edge $(6,7)$. Why does this property hold? precisely because we are sorting nodes co-lexicographically. Take two nodes $u < v$ such that $\lambda(u) = \lambda(v) = a$, for example, $u = 2$ and $v = 7$ (note that $<$ indicates the co-lexicographic order, not pre-order). Since $u < v$, the a labeling edge $(\pi(u), u)$ precedes the a labeling edge $(\pi(v), v)$ in sequence IN, where $\pi(u)$ indicates the parent of u in the tree. In the example, these are the two 'a's appearing at $IN[2]$ and $IN[5]$. Let α_u and α_v be the two strings labeling the two paths from the root to u and v, respectively. In our example, $\alpha_2 = \$a$ and $\alpha_7 = \$ba$ (note that we prepend the artificial incoming label of the root). By the very definition of our co-lexicographic order, $u < v$ if and only if $\alpha_u < \alpha_v$. Note that we can write $\alpha_u = \alpha_{\pi(u)} \cdot a$ and $\alpha_v = \alpha_{\pi(v)} \cdot a$. Then, $\alpha_u < \alpha_v$ holds if and only if $\alpha_{\pi(u)} < \alpha_{\pi(v)}$, i.e., if and only if $\pi(u) < \pi(v)$. In our example, $\alpha_{\pi(2)} = \alpha_1 = \$ < \$b = \alpha_6 = \alpha_{\pi(7)}$; thus, it must hold $1 = \pi(2) < \pi(7) = 6$ (which, in fact, holds in the example). This means that the a labeling edge $(\pi(u), u)$ comes before the a labeling edge $(\pi(v), v)$ also in sequence OUT. In our example, those are the two 'a' contained in $OUT[1]$ and $OUT[6]$. We finally obtain our claim: equally-labeled edges appear in the same relative order in IN and OUT.

The XBWT is a generalization to labeled trees of a well-known string transform—the Burrows-Wheeler transform (BWT) [20]—described for the first time in 1994 (the BWT is precisely sequence OUT of a path tree (To be precise, the original BWT used the symmetric lexicographic order of the string's suffixes). Its corresponding index, the FM-index [16] was first described by Ferragina and Manzini in 2000. The property we just described—allowing the mapping of characters from IN to OUT—is the building block of the FM-index and takes the name *LF mapping*. The LF mapping can be used to perform a visit of the tree using just OUT. First, note that OUT fully specifies IN: the latter is a sorted list of all characters appearing in OUT, plus character $\$$. Start from the virtual incoming edge of the root, which appears always in $IN[1]$. The outgoing labels of the roots appear in $OUT[1]$. Using the LF property, we map the outgoing labels of the root to sequence IN, obtaining the locations in PA of the successors of the root. The reasoning can be iterated, ultimately leading to a complete visit of the tree.

3.5.4. Subpath Queries

The LF mapping can also be used to answer counting queries. Suppose we wish to count the number of nodes reached by a path labeled with string $\Pi[1, m]$. For example, consider the tree of Figure 7 and let $\Pi = aa$. First, find the range $PA[\ell, r]$ of nodes on PA reached by $\Pi[1]$. This is easy, since characters in IN are sorted: $[\ell, r]$ is the maximal range such that $IN[i] = \Pi[1]$ for all $\ell \leq i \leq r$. In our example, $\ell = 2$ and $r = 5$. Now, note that $OUT[\ell, r]$ contains the outgoing labels of all nodes reached by $\Pi[1]$. Then, we can extend our search by one character by following all edges in $OUT[\ell, r]$ labeled with $\Pi[2]$. In our example, $\Pi[2] = a$, and there are 2 edges labeled with 'a' to follow: those at positions $OUT[2]$ and $OUT[3]$. This requires applying the LF mapping to all those edges. Crucially, note that the LF mapping property also guarantees that the nodes we reach by following those edges form a contiguous co-lexicographic range $PA[\ell', r']$. In our example, we obtain the range $PA[3, 4]$, containing pre-order nodes 3 and 4. These are precisely the $occ = \ell' - r + 1$ nodes reached by $\Pi[1, 2]$ (and occ is the answer to the count query for $\Pi[1, 2]$). It is clear that the reasoning can be iterated until finding the range of all nodes reached by a pattern Π of any length.

For clarity, in our discussion above, we have ignored efficiency details. If implemented as described, each extension step would require $O(n)$ time (a linear scan of IN and OUT). It turns out that, using up-to-date data structures [69], a single character-extension step can be implemented in just $O\left(\log\left(\frac{\log \sigma}{\log n}\right)\right)$ time! The idea is, given the range $PA[\ell, r]$ of $\Pi[1]$, to locate just the first and last occurrence of $\Pi[2]$ in $OUT[\ell, r]$. This can be implemented with a so-called *rank* query (that is, the number of characters equal to $\Pi[2]$ before a given position $OUT[i]$). We redirect the curious reader to the original articles by Ferragina et al. [35,68] (XBWT) and Belazzougui and Navarro [69] (up-to-date rank data structures) for the exquisite data structure details.

As far as locate queries are concerned, as previously mentioned a simple strategy could be to explicitly store PA. In the above example, the result of *locate* for pattern $\Pi = aa$ would be the pre-order nodes $PA[3, 4] = 3, 4$. This solution, however, uses $n \log n$ bits on top of the XBWT. Arroyuelo et al. [70] Section 5.1 and Prezza [49] describe a sampling strategy that solves this problem when the nodes' identifiers are DFS numbers: fix a parameter $t \leq n$. The idea is to decompose the tree into $\Theta(n/t)$ subtrees of size $O(t)$, and explicitly store in $O((n/t) \log n)$ bits the pre-order identifiers of the subtrees' roots. Then, the pre-order of any non-sampled node can be retrieved by performing a visit of a subtree using the XBWT navigation primitives. By fixing $t = \log^{1+\epsilon} n$ for any constant $\epsilon > 0$, this strategy can compute any $PA[i]$ (i.e., locate any pattern occurrence) in polylogarithmic time while using just $o(n)$ bits of additional space. A more advanced mechanism [49] allows locating the DFS number of each pattern occurrence in optimal $O(1)$ time within compressed space.

3.5.5. Compression

To conclude, we show how the XBWT (and, similarly, the BWT of a string) can be compressed. In order to efficiently support the LF mapping, sequence IN has to be explicitly stored. However, note that this sequence is strictly increasing. If the alphabet is effective and of the form $\Sigma = [1, \sigma]$, then we can represent IN with a simple bitvector of n bits marking with a bit '1' the first occurrence of each new character. If σ is small (for example, $\text{polylog}(n)$), then this bitvector can be compressed down to $o(n)$ bits while supporting efficient queries [71]. Finally, we concatenate all characters of OUT in a single string and use another bitvector of $2n$ bits to mark the borders between each $OUT[i]$ in the sequence. Still assuming a small alphabet and using up-to-date data structures [69], this representation takes $nH_0 + 2n + o(n)$ bits of space and supports optimal-time count queries. In their original article, Ferragina et al. [35] realized that this space could be further improved: since nodes are sorted in co-lexicographic order, they are also clustered by the paths connecting them to the root. Then, for a sufficiently small $k \in O(\log_\sigma n)$, we can partition OUT by all distinct paths of length k that reach the nodes, and use a different zero-order compressor for each class of the partition. This solution achieves high-order compressed space $nH_k + 2n + o(n)$.

Another method to compress the XBWT is to exploit repetitions of isomorphic subtrees. Alanko et al. [48] show how this can be achieved by a technique they call *tunneling* and that consists in collapsing isomorphic subtrees that are adjacent in co-lexicographic order. Tunneling works for a more general class of graphs (Wheeler graphs), so we discuss it more in detail in Section 3.7. Similarly, one can observe that a repeated topology will generate long runs of equal sets in OUT [49]. Let r be the number of such runs. Repetitive trees (including repetitive texts) satisfy $r \ll n$. In the tree in Figure 7, we have $n = 9$ and $r = 7$. For an example of a more repetitive tree, see Reference [49].

3.6. Further Generalizations

As we have seen, prefix sorting generalizes quite naturally to labeled trees. There is no reason to stop here: trees are just a particular graph topology. A few years after the successful XBWT was introduced, Mantaci et al. [72–74] showed that finite sets of circular strings, i.e., finite collections of disjoint labeled cycles, do enjoy the same *prefix-sortability* property: the nodes of these particular labeled graphs can be arranged by the strings labeling their incoming paths, thus speeding up subsequent substring searches on the graph. Seven years later, Bowe et al. [75] added one more topology to the family: de Bruijn graphs. A de Bruijn graph of order k for a string S (or for a collection of strings; the generalization is straightforward) has one node for each distinct substring of length k (a k-mer) appearing in S. Two nodes, representing k-mers s_1 and s_2, are connected if and only if they overlap by $k - 1$ characters (that is, the suffix of length $k - 1$ of s_1 equals the prefix of the same length of s_2) *and* their concatenation of length $k + 1$ is a substring of S. The topology of a de Bruijn graph could be quite complex; in fact, one could define such a graph starting from the k-mers read on the paths of an arbitrary labeled graph. For large enough k, the set of strings read on the paths of the original graph and the derived de Bruijn graph coincide. Sirén et al. [76] exploited this observation to design a tool to index pan-genomes (that is, genome collections represented as labeled graphs). Given an arbitrary input labeled graph, their GCSA (Generalized Compressed Suffix Array) builds a de Bruijn graph that is equivalent to (i.e., in which paths spell the same strings of) the input graph. While this idea allows to index arbitrary labeled graphs, in the worst case, this conversion could generate an exponentially-larger de Bruijn graph (even though they observe that, in practice, due to the particular distribution of DNA mutations, this exponential explosion does not occur too frequently in bioinformatics applications). A second, more space-efficient version of their tool [77] fixes the maximum order k of the target de Bruijn graph, thus indexing only paths of length at most k of the input graph. Finally, *repeat-free founder block graphs* [78] are yet another recent class of indexable labeled graphs.

In this survey, we do not enter into the details of the above generalizations since they are all particular cases of a larger class of labeled graphs which will be covered in the next subsection: Wheeler graphs. As a result, the search mechanism for Wheeler graphs automatically applies to those particular graph topologies. We redirect the curious reader to the seminal work of Gagie et al. [50] for more details of how these graphs, as well as several other combinatorial objects (FM index of an alignment [79,80], Positional BWT [81], wavelet matrices [82], and wavelet trees [83]), can be elegantly described in the Wheeler graphs framework. In turn, in Section 3.8, we will see that Wheeler graphs are just the "base case" of a larger family of prefix-sortable graphs, ultimately encompassing all labeled graphs: p-sortable graphs.

3.7. Wheeler Graphs

In 2017, Gagie et al. [50] generalized the principle underlying the approaches described in the previous sections to all labeled graphs in which nodes can be sorted co-lexicographically in a total order. They called such objects *Wheeler graphs* in honor of David J. Wheeler, one of the two inventors of the ubiquitous Burrows-Wheeler transform [20] that today stands at the heart of the most successful compressors and compressed text indexes.

Recall that, for convenience, we treat the special (not too restrictive) case of labeled graphs corresponding to the state transition of finite automata. Indexing finite automata allows us to generalize the ideas behind suffix arrays to (possibly infinite) sets of strings: all the strings read from the source to any of the automaton's final states. Said otherwise, an index for a finite automaton is capable of recognizing the substring closure of the underlying regular language. Consider the following regular language:

$$\mathcal{L} = (\epsilon|aa)b(ab|b)^*,$$

as well as consider the automaton recognizing $\mathcal{L}$ depicted in Figure 9.

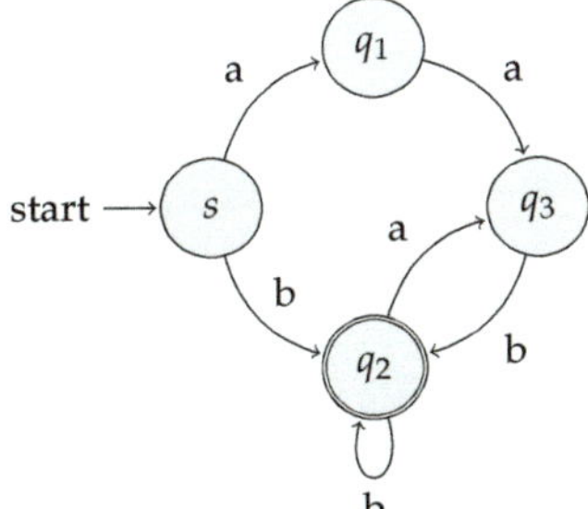

Figure 9. Automaton recognizing the regular language $\mathcal{L} = (\epsilon|aa)b(ab|b)^*$.

Let u be a state, and denote with I_u the (possibly infinite) set of all strings read from the source to u. Following the example reported in Figure 9, we have $I_{q_1} = \{a\}$, $I_{q_2} = \{b, bb, bab, babb, aab, \dots\}$, and $I_{q_3} = \{aa, aaba, aabba, \dots\}$. Intuitively, if the automaton is a DFA, then we are going to sort its states in such a way that two states are placed in the order $u < v$ if and only if all the strings in I_u are co-lexicographically smaller than all the strings in I_v (for NFAs, the condition is slightly more involved as those sets can have a nonempty intersection). Note that, at this point of the discussion, it is not yet clear that such an ordering *always* exists (in fact, we will see that it does not); however, from the previous subsections, we know that particular graph topologies (in particular, paths, cycles, trees, and de Bruijn graphs) do admit a solution to this sorting problem. The legitimate question, tackled for the first time in Gagie et al.'s work, is: what is the largest graph family admitting such an ordering?

Let a, a' be two characters labeling edges (u, u') and (v, v'), respectively. We require the following three *Wheeler properties* [50] for our ordering $\leq$:

(i)   all states with in-degree zero come first in the ordering,
(ii)   if $a < a'$, then $u' < v'$, and
(iii) if $a = a'$ and $u < v$, then $u' \leq v'$.

It is not hard to see that (i)–(iii) generalize the familiar co-lexicographic order among the prefixes of a string to labeled graphs. In this generalized context, we deal with prefixes of the recognized language, i.e., with the strings labeling paths connecting the source state with any other state. Note also that rule (ii) implicitly requires that all labels entering a state must be equal. This is not a severe restriction: at the cost of increasing the number of states by a factor σ, any automaton can be transformed into an equivalent one with this property [65]. Figure 10 depicts the automaton of Figure 9 after having ordered (left-to-right) the states according to the above three Wheeler properties.

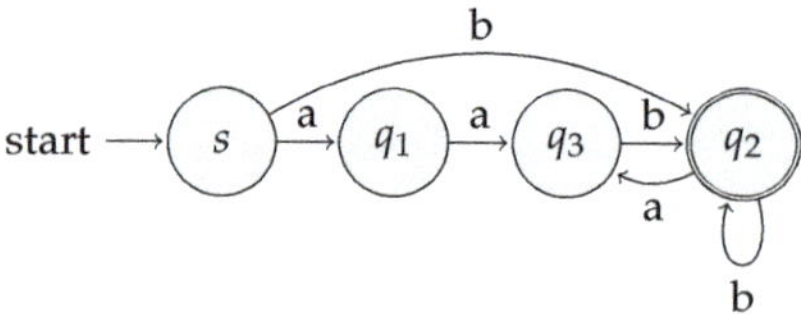

Figure 10. The automaton of Figure 9, sorted according to the three Wheeler properties.

An ordering $\leq$ satisfying Wheeler properties (i)–(iii) is called a *Wheeler order*. Note that a Wheeler order, when it exists, is always total (this will become important in Section 3.8). An automaton (or a labeled graph) is said to be Wheeler if its states admit at least one Wheeler order.

3.7.1. Subpath Queries

When a Wheeler order exists, all the power of suffix arrays can be lifted to labeled graphs: (1) the nodes reached by a path labeled with a given pattern Π form a consecutive range in Wheeler order, and (2) such a range can be found in linear time as a function of the pattern's length. In fact, the search algorithm generalizes those devised for the particular graphs discussed in the previous sections. Figure 11 depicts the process of searching all nodes reached by a path labeled with pattern $\Pi = $ "aba". The algorithm starts with $\Pi = \epsilon$ (empty string, all nodes) and right-extends it by one letter step by step, following the edges labeled with the corresponding character of Π from the current set of nodes. Note the following crucial property: at each step, the nodes reached by the current pattern form a *consecutive range*. This makes it possible to represent the range in constant space (by specifying just the first and last node in the range). Similarly to what we briefly discussed in Section 3.5 (*Subpath Queries on the XBWT*), by using appropriate compressed data structures supporting rank and select on strings, each character extension can be implemented efficiently (i.e., in log-logarithmic time). The resulting data structure can be stored in entropy-compressed space [50].

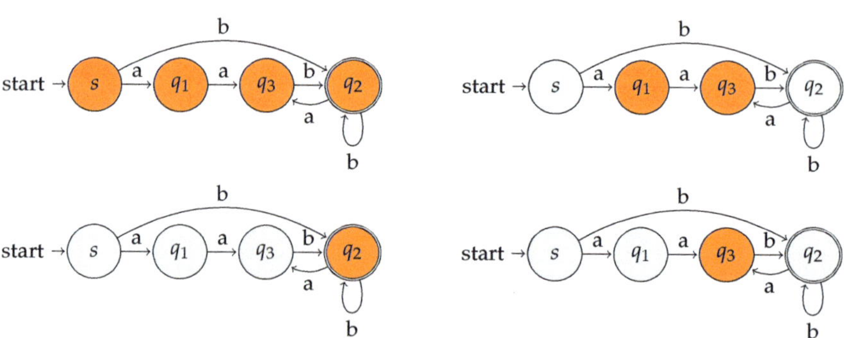

Figure 11. Searching nodes reached by a path labeled "aba" in a Wheeler graph. Top left: we begin with the nodes reached by the empty string (full range). Top right: range obtained from the previous one following edges labeled 'a'. Bottom left: range obtained from the previous one following edges labeled 'b'. Bottom right: range obtained from the previous one following edges labeled 'a'. This last range contains all nodes reached by a path labeled "aba"

The algorithm we just described allows us to find the number of nodes reached by a given pattern, i.e., to solve *count* queries on the graph. As far as *locate* queries are concerned, Sirén et al. in Reference [76] proposed a sampling scheme that returns the labels of all nodes reached by the query pattern, provided that the labeling scheme satisfies a particular monotonicity property (that is, the labels of a node are larger than those of its predecessors). In practical situations, such as genome-graph indexing, such labels can be equal to an absolute position on a reference genome representing all the paths of the graph. In the worst case, however, this scheme requires to store $\Theta(n)$ labels of $O(\log n)$ bits each. As another option, the sampling strategy described by Arroyuelo et al. [70] Section 5.1 and Prezza [49] (see Section 3.5.4 for more details) can be directly applied to the DFS spanning forest of any Wheeler graph, thus yielding locate queries in polylogarithmic time and small additiona redundancy.

3.7.2. Compression

The nodes of a Wheeler graph are sorted (clustered) with respect to their incoming paths, so the outgoing labels of adjacent nodes are likely to be similar. This allows applying high-order compression to the labels of a Wheeler graph [50]. It is even possible to compress the graph's topology (together with the labels), if this is highly repetitive (i.e., the graph has large repeated isomorphic subgraphs). By generalizing the *tunneling* technique originally devised by Baier [84] for the Burrows-Wheeler transform, Alanko et al. [48] showed that isomorphic subtrees adjacent in co-lexicographic order can be collapsed while maintaining the Wheeler properties. In the same paper, they showed that a *tunneled* Wheeler graph can be even indexed to support existence path queries (a relaxation of counting queries where we can only discover whether or not a query pattern labels some path in the graph). Similarly, one can observe that a repetitive graph topology will generate long runs of equal sets of outgoing labels in Wheeler order. This allows applying run-length compression to the Wheeler graph [49]. Run-length compression of a Wheeler graph is equivalent to collapsing isomorphic subgraphs that are adjacent in Wheeler order and, thus, shares many similarities with the tunneling technique. A weaker form of run-length compression of a Wheeler graph has been considered by Bentley et al. in Reference [85]. In this work, they turn a Wheeler graph into a string by concatenating the outgoing labels of the sorted nodes, permuted so as to minimize the number of equal-letter runs of the resulting string. They show that the (decision version of the) problem of finding the ordering of the Wheeler graph's sources that minimizes the number of runs is NP-complete. In the same work, they show that also the problem of finding the alphabet ordering minimizing the number of equal-letter runs in the BWT is NP-complete.

3.7.3. Sorting and Recognizing Wheeler Graphs

Perhaps not surprisingly, not all labeled graphs admit a Wheeler order of their nodes. Intuitively, this happens because of conflicting predecessors: if u and v have a pair of predecessors ordered as $u' < v'$ and another pair ordered as $v'' < u''$, then a Wheeler order cannot exist. As an example, consider any automaton recognizing the language $\mathcal{L}' = (ax^*b)|(cx^*d)$ (original example by Gagie et al. [50]). Since any string beginning with letter 'a' must necessarily end with letter 'b' and, similarly, any string beginning with letter 'c' must necessarily end with letter 'd', the paths spelling ax^kb and $ax^{k'}d$ must be disjoint for all $k, k' \geq 0$. Said otherwise, the nodes reached by label 'x' and leading to 'b' must be disjoint from those reached by label 'x' and leading to 'd'. Denote with u_α a state reached by string α (read from the source). Then, it must be $u_{ax} < u_{bx}$: from the above observation, those two states must be distinct, and, from the Wheeler properties, they must be in this precise order. For the same reason, it must be the case that $u_{bx} < u_{axx} < u_{bxx} < u_{axxx} < \cdots$. This is an infinite sequence of distinct states: no finite Wheeler automaton can recognize $\mathcal{L}'$.

This motivates the following natural questions: given an automaton $\mathcal{A}$, is it Wheeler? if yes, can we efficiently find a corresponding Wheeler order? It turns out that these are hard problems. Gibney and Thankachan showed in Reference [86] that deciding whether an arbitrary automaton admits a Wheeler order is NP-complete. This holds even when each state is allowed to have at most five outgoing edges labeled with the same character (which somehow bounds the amount of nondeterminism). On the positive side, Alanko et al. [65]

showed that both problems (recognition and sorting) can be solved in quadratic time when each state is allowed to have at most two outgoing edges labeled with the same character. This includes DFAs, for which, however, a more efficient linear-time solution exists [65]. The complexity of the cases in between, i.e., at most three/four equally-labeled outgoing edges, is still open [87].

3.7.4. Wheeler Languages

Since we are talking about finite automata, one question should arise naturally: what languages are recognized by Wheeler NFAs? Let us call *Wheeler languages* this class. First of all, Wheeler languages are clearly regular since, by definition, they are accepted by finite state automata. Moreover, all finite languages are Wheeler because they can be recognized by a tree-shaped automaton, which (as seen in Section 3.5) is always prefix-sortable. Additionally, as observed by Gagie et al. [50] not all regular languages are Wheeler: $(ax^*b)|(cx^*d)$ is an example. Interestingly, Wheeler languages can be defined both in terms of DFAs and NFAs: Alanko et al. proved in Reference [65] that Wheeler NFAs and Wheeler DFAs recognize the same class of languages. Another powerful language-theoretic result that can be transferred from regular to Wheeler languages is their neat algebraic characterization based on Myhill-Nerode equivalence classes [66]. We recall that two strings α and β are Myhill-Nerode equivalent with respect to a given regular language $\mathcal{L}$ if and only if, for any string γ, we have that $\alpha\gamma \in \mathcal{L} \Leftrightarrow \beta\gamma \in \mathcal{L}$. The Myhill-Nerode theorem states that $\mathcal{L}$ is regular if and only if the Myhill-Nerode equivalence relation has finite index (i.e., it has a finite number of equivalence classes). In the Wheeler case, the Myhill-Nerode equivalence relation is slightly modified by requiring that equivalence classes of prefixes of the language are also intervals in co-lexicographic order and contain words ending with the same letter. After this modification, the Myhill-Nerode theorem can be transferred to Wheeler languages: $\mathcal{L}$ is Wheeler if and only if the modified Myhill-Nerode equivalence relation has finite index [65,88]. See Alanko et al. [88] for a comprehensive study of the elegant properties of Wheeler languages (including closure properties and complexity of the recognition problem). From the algorithmic point of view (which is the most relevant to this survey), these results permit to define and build efficiently the *minimum* Wheeler DFA (that is, with the smallest number of states) recognizing the same language of a given input Wheeler DFA, thus optimizing the space of the resulting index [65].

After this introduction to Wheeler languages, let us consider the question: can we index Wheeler languages efficiently? Interestingly, Gibney and Thankachan's NP-completeness proof [86] requires that the automaton's topology is fixed so it does not rule out the possibility that we can index in polynomial time an equivalent automaton. After all, in many situations, we are actually interested in indexing a language rather than a fixed graph topology. Surprisingly, the answer to the above question is yes: it is easier to index Wheeler languages, rather than Wheeler automata. Since recognizing and sorting Wheeler DFAs is an easy problem, a first idea could be to turn the input NFA into an equivalent DFA. While it is well-known that, in the worst case, this conversion (via the powerset algorithm) could result in an exponential blow-up of the number of states, Alanko et al. [65] proved that a Wheeler NFA always admits an equivalent Wheeler DFA of linear size (the number of states doubles at most). This has an interesting consequence: if the input NFA is Wheeler, then we can index its language in linear time (in the size of the input NFA) after converting it to a DFA (which requires polynomial time). We can actually do better: if the input NFA $\mathcal{A}$ is Wheeler, then in polynomial time we can build a Wheeler NFA $\mathcal{A}'$ that (i) is never larger than $\mathcal{A}$, (ii) recognizes the same language as $\mathcal{A}$, and (iii) can be sorted in polynomial time [88]. We remark that there is a subtle reason why the above two procedures for indexing Wheeler NFAs do not break the NP-completeness of the problem of recognizing this class of graphs: it is possible that they generate a Wheeler NFA even if the input is not a Wheeler NFA; thus they cannot be used to solve the recognition problem. To conclude, these strategies can be used to index Wheeler NFAs, but do not tell us anything about indexing Wheeler languages represented as a general (possibly non-Wheeler) NFA. An interesting case is represented by finite languages (always Wheeler) represented as acyclic NFAs. The lower bounds discussed in Section 3.2 tell us that a quadratic running time is unavoidable for the graph indexing problem, even in the acyclic case. This implies, in particular, that the conversion from arbitrary acyclic NFAs to Wheeler NFAs must incur in a quadratic blow-up in the worst case. In practice, the situation is much worse: Alanko et al. showed in Reference [65] that the blow-up is exponential in the worst case. In the same paper, they provided a fast algorithm to convert any acyclic DFA into the smallest equivalent Wheeler DFA.

3.8. p-Sortable Automata

Despite its power, the Wheeler graph framework does not allow to index and compress *any* labeled graph: it is easy to come up with arbitrarily large Wheeler graphs that lose their Wheeler property after the addition of just *one* edge. Does this mean that such an augmented graph cannot be indexed efficiently? clearly not, since it would be sufficient to add a small "patch" (keeping track of the extra edge) to the index of the underlying Wheeler subgraph in order to index it. As another "unsatisfying" example, consider the union $\mathcal{L} = \mathcal{L}_1 \cup \mathcal{L}_2$ of two Wheeler languages $\mathcal{L}_1$ and $\mathcal{L}_2$. In general, $\mathcal{L}$ is not a Wheeler language [88]. However, $\mathcal{L}$ can be easily indexed by just keeping two indexes (of two Wheeler automata recognizing $\mathcal{L}_1$ and $\mathcal{L}_2$), *with no asymptotic slowdown* in query times! The latter example is on the right path to a solution of the problem. Take two Wheeler automata $\mathcal{A}_1$ and $\mathcal{A}_2$ recognizing two Wheeler languages $\mathcal{L}_1$ and $\mathcal{L}_2$, respectively, such that $\mathcal{L} = \mathcal{L}_1 \cup \mathcal{L}_2$ is not Wheeler. The union automaton $\mathcal{A}_1 \cup \mathcal{A}_2$ (obtained by simply merging the start states of $\mathcal{A}_1$ and $\mathcal{A}_2$) is a nondeterministic automaton recognizing $\mathcal{L}$. Taken individually, the states of $\mathcal{A}_1$ and $\mathcal{A}_2$ can be sorted in two total orders. However, taken as a whole, the two sets of states do not admit a total co-lexicographic order (which would imply that $\mathcal{L}$ is Wheeler). The solution to this riddle is to abandon total orders in favor of *partial orders* [89]. A partial order $\leq$ on a set V (in our case, the set of the automaton's states) is a reflexive, antisymmetric and transitive relation on V. In a partial order, two elements u, v either are comparable, in which case $u \leq v$ or $v \leq u$ hold (both hold only if $u = v$), or are not comparable, in which case neither $u \leq v$ nor $v \leq u$ hold. The latter case is indicated as $u \parallel v$. Our co-lexicographic partial order is defined as follows. Let a, a' be two characters labeling edges (u, u') and (v, v'), respectively. We require the following properties:

(i)   all states with in-degree zero come first in the ordering,
(ii)   if $a < a'$, then $u' < v'$, and
(iii) if $a = a'$ and $u' < v'$, then $u \leq v$.

Note that, differently from the definition of Wheeler order (Section 3.7), the implication of (iii) follows the edges *backwards*. As it turns out, $\leq$ is indeed a partial order and, as we show in the next subsections, it allows generalizing the useful properties of Wheeler graphs to arbitrary topologies. A convenient representation for any partial order is a Hasse diagram: a directed acyclic graph where we draw the elements of the order from the smallest (bottom) to largest ones (top), and two elements are connected by an edge (u, v) if and only if $u \leq v$. See Figure 12 for an example.

The lower bounds discussed in Section 3.2 tell us that indexed pattern matching on graphs cannot be solved faster than quadratic time in the worst case. In particular, this means that our generalization of Wheeler graphs cannot yield an index answering subpath queries in linear time. In fact, there is a catch: the extent to which indexing and compression can be performed efficiently is proportional to the similarity of the partial order $\leq$ to a total one. As it turns out, the correct measure of similarity to consider is the *order's width*: the minimum number p of totally-ordered chains (subsets of states) into which the set of states can be partitioned. Figure 12 makes it clear that the states of the automaton can be divided into $p = 2$ totally-ordered subsets. This choice is not unique (look at the Hasse diagram), and a possible partition is $s < q_1 < q_3$ (in yellow) and q_2 (in red). We call p-sortable the class of automata for which there exists a chain partition of size p. Since the

n states of any automaton can always be partitioned into n chains, this definition captures all automata. The parameter p seems to capture some deep regularity of finite automata: in addition to determining the compressibility of their topology (read next subsections), it also determines their inherent determinism: a p-sortable NFA with n states always admits an equivalent DFA (which can be obtained via the standard powerset algorithm) with at most $2^p(n - p + 1) - 1$ states. This represents some sort of fine-grained analysis refining the folklore (tight) bound of 2^n [90] and has deep implications to several algorithms on automata. For example, it implies that the PSPACE-complete NFA equivalence problem is fixed-parameter tractable with respect to p [89]. To conclude, we mention that finding the smallest p for a given labeled graph is NP complete, though the problem admits an $O(e^2 + n^{5/2})$-time algorithm for the deterministic case [89].

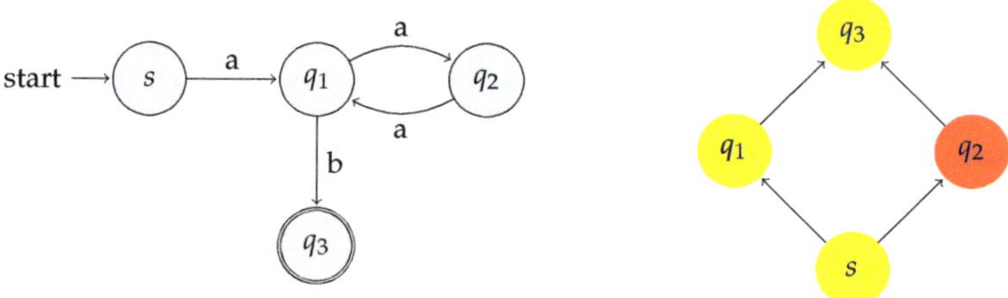

Figure 12. (**Left**) Automaton for the language $\mathcal{L} = a(aa)^*b$. (**Right**) Hasse diagram of a co-lexicographic partial order for the graph. The order's width is 2, and the order can be partitioned into two totally-ordered chains (yellow and red in the example; this is not the only possible choice).

3.8.1. Subpath Queries

Subpath queries can be answered on p-sortable automata by generalizing the forward search algorithm of Wheeler graphs. In fact, the following property holds: for any pattern Π, the states reached by a path labeled with Π always form one convex set in the partial order. In turn, any convex set can be expressed as p intervals, one contained in each class of the chain partition [89] (for any chain partition). The generalized forward search algorithm works as follows: start with the interval of ϵ on the p chains (the full interval on each chain). For each character a of the pattern, follow the edges labeled with a that depart from the current (at most) p intervals. By the above property, the target nodes will still be contained in (at most) p intervals on the chains. Consider the example of Figure 12. In order to find all nodes reached by pattern 'aa', the generalized forward search algorithm first finds the nodes reached by 'a': those are nodes q_1 (a range on the yellow chain) and q_2 (a range on the red chain). Finally, the algorithm follows all edges labeled 'a' from those ranges, thereby finding all nodes reached by 'aa': nodes (again) q_1 and q_2 which, as seen above, form one range on the yellow chain and one on the red chain. See [89] for a slightly more involved example. Using up-to-date data structures, the search algorithm can be implemented to run in $O(|\Pi| \cdot \log(\sigma p) \cdot p^2)$ steps [89]. In the worst case ($p = n$), this essentially matches the lower bound of Equi et al. [52–54] on dense graphs (see Section 3.2). On the other hand, for small values of p this running time can be significantly smaller than the lower bound: in the extreme case, $p = 1$, and we obtain the familiar class of Wheeler graphs (supporting optimal-time subpath queries).

As far as locate queries are concerned, the sampling strategy described by Arroyuelo et al. [70] Section 5.1 and Prezza [49] (see Section 3.5.4 for more details) can be directly applied to the DFS spanning forest of any p-sortable graph (similarly to Wheeler graphs), thus yielding locate queries in polylogarithmic time and small additional redundancy. This is possible since the index for p-sortable graphs described in Reference [89] supports navigation primitives, as well, and can thus be used to navigate the DFS spanning forest of the graph.

3.8.2. Compression

The Burrows-Wheeler transform [20] (also see Section 3.5) can be generalized to p-sortable automata: the main idea is to (i) partition the states into p totally-sorted chains (for example, in Figure 12, one possible such chain decomposition is $\{s, q_1, q_3\}, \{q_2\}$), (ii) order the states by "gluing" the chains in any order (for example, in Figure 12 one possible such ordering is s, q_1, q_3, q_2), (iii) build the adjacency matrix of the graph using this state ordering and (iv) for each edge in the matrix, store only its label and its endpoint chains (that is, two numbers between 1 and p indicating which chains the two edges' endpoints belong to). It can be shown that this matrix can be linearized in an invertible representation taking $\log \sigma + 2 \log p + 2 + o(1)$ bits per edge. On DFAs, the representation takes less space: $\log \sigma + \log p + 2 + o(1)$ bits per edge [89]. This is already a compressed representation of the graph, since "sortable" graphs (i.e., having small p) are compressed to few bits per edge, well below the information-theoretic worst-case lower bound if $p \ll n$. Furthermore, within each chain the states are sorted by their incoming paths. It follows that, as seen in Sections 3.5 and 3.7, high-order entropy-compression can be applied within each chain.

4. Conclusions and Future Challenges

In this survey, we tried to convey the core ideas that have been developed to date in the field of compressed graph indexing, in the hope of introducing (and attracting) the non-expert reader to this exciting and active area of research. As it can be understood by reading our survey, the field is in rapid expansion and does not lack of stimulating challenges. On the lower-bound side, a possible improvement could be to provide fine-grained lower bounds to the graph indexing problem, e.g., as a function of the parameter p introduced in Section 3.8. As far as graph compression is concerned, it is plausible that new powerful graph compression schemes will emerge from developments in the field: an extension of the run-length Burrows Wheeler transform to labeled graphs, for example, could compete with existing grammar-based graph compressors while also supporting efficient path queries. Efficient index construction is also a considerable challenge. As seen in Section 3.7, Wheeler graphs can be sorted efficiently only in the deterministic case. Even when the nondeterminism degree is very limited to just two equally-labeled edges leaving a node, the fastest sorting algorithm has quadratic complexity. Even worse, in the general (nondeterministic) case, the problem is NP-complete. The situation does not improve with the generalization considered in Section 3.8: finding the minimum width p for a deterministic graph takes super-quadratic time with current solutions, and it becomes NP-complete for arbitrary graphs. Clearly, practical algorithms for graph indexing will have to somehow sidestep these issues. One possible solution could be approximation: we note that, even if the optimal p cannot be found efficiently (recall that $p = 1$ for Wheeler graphs), approximating it (for example, up to a polylog(n) factor) would still guarantee fast queries. Further challenges include generalizing the labeled graph indexing problem to allow aligning regular expressions against graphs. This could be achieved, for example, by indexing both the graph and an NFA recognizing the query (regular expression) pattern.

Funding: This research received no external funding.

Institutional Review Board Statement: Not applicable.

Informed Consent Statement: Not applicable.

Data Availability Statement: Data sharing not applicable.

Acknowledgments: I would like to thank Alberto Policriti for reading a preliminary version of this manuscript and providing useful suggestions.

Abbreviations

The following abbreviations are used in this manuscript:

BWT	Burrows-Wheeler Transform
DFA	Deterministic Finite Automaton
GCSA	Generalized Compressed Suffix Array
NFA	Nondeterministic Finite Automaton
PA	Prefix array
SA	Suffix Array
XBWT	eXtended Burrows-Wheeler Transform
$\mathcal{T}$	Text (string to be indexed)
Π	Pattern to be aligned on the labeled graph
Σ	Alphabet
σ	Size of the alphabet: $\sigma = |\Sigma|$
e	Number of edges in a labeled graph
m	Pattern length: $m = |\Pi|$
n	Length of a text or number of nodes in a labeled graph
p	Number of chains in a chain decomposition of the automaton's partial order (Section 3.8)

References

1. Karp, R.M.; Rabin, M.O. Efficient randomized pattern-matching algorithms. *IBM J. Res. Dev.* **1987**, *31*, 249–260. [CrossRef]
2. Galil, Z. On improving the worst case running time of the Boyer-Moore string matching algorithm. *Commun. ACM* **1979**, *22*, 505–508. [CrossRef]
3. Apostolico, A.; Giancarlo, R. The Boyer-Moore-Galil string searching strategies revisited. *SIAM J. Comput.* **1986**, *15*, 98–105. [CrossRef]
4. Knuth, D.E.; Morris, J.H., Jr.; Pratt, V.R. Fast Pattern Matching in Strings. *SIAM J. Comput.* **1977**, *6*, 323–350. [CrossRef]
5. Navarro, G. *Compact Data Structures—A Practical Approach*; Cambridge University Press: Cambridge, UK, 2016; p. 536, ISBN 978-1-107-15238-0.
6. Mäkinen, V.; Belazzougui, D.; Cunial, F.; Tomescu, A.I. *Genome-Scale Algorithm Design*; Cambridge University Press: Cambridge, UK, 2015.
7. Navarro, G.; Mäkinen, V. Compressed Full-Text Indexes. *ACM Comput. Surv.* **2007**, *39*, 2. [CrossRef]
8. Navarro, G. Indexing Highly Repetitive String Collections, Part I: Repetitiveness Measures. Available online: https://arxiv.org/abs/2004.02781 (accessed on 10 November 2020).
9. Navarro, G. Indexing Highly Repetitive String Collections, Part II: Compressed Indexes. Available online: https://link.springer.com/chapter/10.1007/978-3-642-35926-2_29 (accessed on 10 November 2020).
10. The Computational Pan-Genomics Consortium. Computational pan-genomics: Status, promises and challenges. *Briefings Bioinform.* **2016**, *19*, 118–135. [CrossRef]
11. Manber, U.; Myers, G. Suffix arrays: A new method for on-line string searches. In Proceedings of the First Annual ACM-SIAM Symposium on Discrete Algorithms, San Francisco, CA, USA, 22–24 January 1990; pp. 319–327.
12. Baeza-Yates, R.A.; Gonnet, G.H. A New Approach to Text Searching. In Proceedings of the 12th Annual International ACM SIGIR Conference on Research and Development in Information Retrieval (SIGIR'89), Cambridge, MA, USA, 25–28 June 1989; ACM: New York, NY, USA, 1989; pp. 168–175. [CrossRef]
13. Gonnet, G.H.; Baeza-Yates, R.A.; Snider, T. New Indices for Text: Pat Trees and Pat Arrays. *Inf. Retr. Data Struct. Algorithms* **1992**, *66*, 82.
14. Weiner, P. Linear pattern matching algorithms. In Proceedings of the 14th Annual Symposium on Switching and Automata Theory (swat 1973), Iowa City, IA, USA, 15–17 October 1973; pp. 1–11.
15. Kärkkäinen, J.; Ukkonen, E. Lempel-Ziv parsing and sublinear-size index structures for string matching. In Proceedings of the 3rd South American Workshop on String Processing (WSP'96), Recife, Brazil, 8–9 August 1996.
16. Ferragina, P.; Manzini, G. Opportunistic data structures with applications. In Proceedings of the 41st Annual Symposium on Foundations of Computer Science, Redondo Beach, CA, USA, 12–14 November 2000; pp. 390–398.
17. Grossi, R.; Vitter, J.S. Compressed Suffix Arrays and Suffix Trees with Applications to Text Indexing and String Matching (Extended Abstract). In Proceedings of the Thirty-Second Annual ACM Symposium on Theory of Computing (STOC'00), Portland, OR, USA, 21–23 May 2000; Association for Computing Machinery: New York, NY, USA, 2000; pp. 397–406. [CrossRef]
18. Elias, P. Efficient Storage and Retrieval by Content and Address of Static Files. *J. ACM* **1974**, *21*, 246–260. [CrossRef]
19. Fano, R.M. *On the Number of Bits Required to Implement an Associative Memory*; Project MAC; Massachusetts Institute of Technology: Cambridge, MA, USA, 1971.
20. Burrows, M.; Wheeler, D.J. A Block-Sorting Lossless Data Compression Algorithm. Available online: http://citeseerx.ist.psu.edu/viewdoc/summary?doi=10.1.1.3.8069 (accessed on 10 November 2020).

21. Langmead, B.; Trapnell, C.; Pop, M.; Salzberg, S.L. Ultrafast and memory-efficient alignment of short DNA sequences to the human genome. *Genome Biol.* **2009**, *10*, R25. [CrossRef]

22. Li, H.; Durbin, R. Fast and accurate short read alignment with Burrows—Wheeler transform. *Bioinformatics* **2009**, *25*, 1754–1760. [CrossRef]

23. Kreft, S.; Navarro, G. On Compressing and Indexing Repetitive Sequences. *Theor. Comput. Sci.* **2013**, *483*, 115–133. [CrossRef]

24. Gagie, T.; Navarro, G.; Prezza, N. Fully Functional Suffix Trees and Optimal Text Searching in BWT-Runs Bounded Space. *J. ACM* **2020**, *67*. [CrossRef]

25. Mäkinen, V.; Navarro, G. Succinct suffix arrays based on run-length encoding. In Proceedings of the Annual Symposium on Combinatorial Pattern Matching, Jeju Island, Korea, 19–22 June 2005.

26. Sirén, J.; Välimäki, N.; Mäkinen, V.; Navarro, G. Run-Length Compressed Indexes Are Superior for Highly Repetitive Sequence Collections. In Proceedings of the 15th International Symposium on String Processing and Information Retrieval (SPIRE), Melbourne, Australia, 10–12 November 2008; pp. 164–175.

27. Claude, F.; Navarro, G. Improved Grammar-Based Compressed Indexes. In Proceedings of the 19th International Symposium on String Processing and Information Retrieval (SPIRE), Cartagena de Indias, Colombia, 21–25 October 2012; pp. 180–192.

28. Navarro, G.; Prezza, N. Universal compressed text indexing. *Theor. Comput. Sci.* **2019**, *762*, 41–50. [CrossRef]

29. Kempa, D.; Prezza, N. At the Roots of Dictionary Compression: String Attractors. In Proceedings of the 50th Annual ACM SIGACT Symposium on Theory of Computing (STOC 2018), Los Angeles, CA, USA, 25–29 June 2018; Association for Computing Machinery: New York, NY, USA, 2018; pp. 827–840. [CrossRef]

30. Kociumaka, T.; Navarro, G.; Prezza, N. Towards a Definitive Measure of Repetitiveness. In Proceedings of the 14th Latin American Symposium on Theoretical Informatics (LATIN), Sao Paulo, Brazil, 25–19 May 2020; to appear.

31. Gagie, T.; Navarro, G.; Prezza, N. On the approximation ratio of Lempel-Ziv parsing. In Proceedings of the Latin American Symposium on Theoretical Informatics, Buenos Aires, Argentina, 14–19 April 2018.

32. Christiansen, A.R.; Ettienne, M.B.; Kociumaka, T.; Navarro, G.; Prezza, N. Optimal-Time Dictionary-Compressed Indexes. *ACM Trans. Algorithms* **2020**, *31*, 1–39. [CrossRef]

33. Maneth, S.; Peternek, F. A Survey on Methods and Systems for Graph Compression. *arXiv* **2015**, arXiv:1504.00616.

34. Besta, M.; Hoefler, T. Survey and Taxonomy of Lossless Graph Compression and Space-Efficient Graph Representations. *arXiv* **2019**, arXiv:1806.01799.

35. Ferragina, P.; Luccio, F.; Manzini, G.; Muthukrishnan, S. Compressing and Indexing Labeled Trees, with Applications. *J. ACM* **2009**, *57*, 1–33. [CrossRef]

36. Ferres, L.; Fuentes-Sepúlveda, J.; Gagie, T.; He, M.; Navarro, G. Fast and Compact Planar Embeddings. *Comput. Geom. Theory Appl.* **2020**, *89*, 101630. [CrossRef]

37. Chakraborty, S.; Grossi, R.; Sadakane, K.; Rao Satti, S. Succinct Representation for (Non) Deterministic Finite Automata. *arXiv* **2019**, arXiv:1907.09271.

38. Brisaboa, N.; Ladra, S.; Navarro, G. Compact Representation of Web Graphs with Extended Functionality. *Inf. Syst.* **2014**, *39*, 152–174. [CrossRef]

39. Jansson, J.; Sadakane, K.; Sung, W.K. Ultra-succinct representation of ordered trees with applications. *J. Comput. Syst. Sci.* **2012**, *78*, 619–631. [CrossRef]

40. Hucke, D.; Lohrey, M.; Benkner, L.S. Entropy Bounds for Grammar-Based Tree Compressors. In Proceedings of the 2019 IEEE International Symposium on Information Theory (ISIT), Paris, France, 7–12 July 2019; pp. 1687–1691.

41. Gańczorz, M. Using Statistical Encoding to Achieve Tree Succinctness Never Seen Before. In Proceedings of the 37th International Symposium on Theoretical Aspects of Computer Science (STACS 2020), Montpellier, France, 10–13 March 2020.

42. Hucke, D.; Lohrey, M.; Benkner, L.S. A Comparison of Empirical Tree Entropies. In *String Processing and Information Retrieval*; Boucher, C., Thankachan, S.V., Eds.; Springer International Publishing: Cham, Switzerland, 2020; pp. 232–246.

43. Engelfriet, J. Context-Free Graph Grammars. In *Handbook of Formal Languages: Volume 3 Beyond Words*; Rozenberg, G., Salomaa, A., Eds.; Springer: Berlin/Heidelberg, Germany, 1997; pp. 125–213. [CrossRef]

44. Maneth, S.; Peternek, F. Grammar-based graph compression. *Inf. Syst.* **2018**, *76*, 19–45. [CrossRef]

45. Maneth, S.; Peternek, F. Constant delay traversal of grammar-compressed graphs with bounded rank. *Inf. Comput.* **2020**, *273*, 104520. [CrossRef]

46. Gawrychowski, P.; Jez, A. LZ77 factorisation of trees. In Proceedings of the 36th IARCS Annual Conference on Foundations of Software Technology and Theoretical Computer Science (FSTTCS 2016), Madras, India, 15–17 December 2016.

47. Bille, P.; Gørtz, I.L.; Landau, G.M.; Weimann, O. Tree Compression with Top Trees. In *Automata, Languages, and Programming*; Fomin, F.V., Freivalds, R., Kwiatkowska, M., Peleg, D., Eds.; Springer: Berlin/Heidelberg, Germany, 2013; pp. 160–171.

48. Alanko, J.; Gagie, T.; Navarro, G.; Seelbach Benkner, L. Tunneling on Wheeler Graphs. In Proceedings of the 2019 Data Compression Conference (DCC), Snowbird, UT, USA, 26–29 March 2019; pp. 122–131. [CrossRef]

49. Prezza, N. On Locating Paths in Compressed Tries. In Proceedings of the 2021 ACM-SIAM Symposium on Discrete Algorithms (SODA'21), Alexandria, VA, USA, 10–13 January 2021; ACM, Society for Industrial and Applied Mathematics: Philadelphia, PA, USA, 2021.

50. Gagie, T.; Manzini, G.; Sirén, J. Wheeler graphs: A framework for BWT-based data structures. *Theor. Comput. Sci.* **2017**, *698*, 67–78. [CrossRef] [PubMed]

51. Backurs, A.; Indyk, P. Which regular expression patterns are hard to match? In Proceedings of the 2016 IEEE 57th Annual Symposium on Foundations of Computer Science (FOCS), New Brunswick, NJ, USA, 9–11 October 2016; pp. 457–466.

52. Equi, M.; Mäkinen, V.; Tomescu, A.I. Conditional Indexing Lower Bounds Through Self-Reducibility. *arXiv* **2020**, arXiv:2002.00629.

53. Equi, M.; Mäkinen, V.; Tomescu, A.I. Graphs cannot be indexed in polynomial time for sub-quadratic time string matching, unless SETH fails. *arXiv* **2020**, arXiv:2002.00629.

54. Equi, M.; Grossi, R.; Mäkinen, V.; Tomescu, A.I. On the Complexity of String Matching for Graphs. In Proceedings of the 46th International Colloquium on Automata, Languages, and Programming (ICALP 2019), Patras, Greece, 9–12 July 2019. [CrossRef]

55. Impagliazzo, R.; Paturi, R. On the Complexity of K-SAT. *J. Comput. Syst. Sci.* **2001**, *62*, 367–375. [CrossRef]

56. Potechin, A.; Shallit, J. Lengths of words accepted by nondeterministic finite automata. *Inf. Process. Lett.* **2020**, *162*, 105993. [CrossRef]

57. Williams, R. A New Algorithm for Optimal 2-Constraint Satisfaction and Its Implications. *Theor. Comput. Sci.* **2005**, *348*, 357–365. [CrossRef]

58. Gibney, D.; Hoppenworth, G.; Thankachan, S.V. Simple Reductions from Formula-SAT to Pattern Matching on Labeled Graphs and Subtree Isomorphism. *arXiv* **2020**, arXiv:2008.11786.

59. Abboud, A.; Bringmann, K. Tighter Connections Between Formula-SAT and Shaving Logs. In *Proceedings of the 45th International Colloquium on Automata, Languages, and Programming (ICALP 2018), Prague, Czech Republic, 9–13 July 2018*; Chatzigiannakis, I., Kaklamanis, C., Marx, D., Sannella, D., Eds.; Schloss Dagstuhl–Leibniz-Zentrum fuer Informatik: Dagstuhl, Germany, 2018; Volume 107, pp. 8:1–8:18. [CrossRef]

60. Amir, A.; Lewenstein, M.; Lewenstein, N. Pattern matching in hypertext. *J. Algorithms* **2000**, *35*, 82–99. [CrossRef]

61. Ferragina, P.; Mishra, B. Algorithms in Stringomics (I): Pattern-Matching against "Stringomes". *BioRxiv* **2014**, 001669. [CrossRef]

62. Thachuk, C. Indexing hypertext. *J. Discret. Algorithms* **2013**, *18*, 113–122. [CrossRef]

63. Manber, U.; Wu, S. Approximate String Matching with Arbitrary Costs for Text and Hypertext. Available online: https://www.worldscientific.com/doi/abs/10.1142/9789812797919_0002 (accessed on 10 November 2020).

64. Navarro, G. Improved approximate pattern matching on hypertext. In *LATIN'98: Theoretical Informatics*; Lucchesi, C.L., Moura, A.V., Eds.; Springer: Berlin/Heidelberg, Germany, 1998; pp. 352–357.

65. Alanko, J.; D'Agostino, G.; Policriti, A.; Prezza, N. Regular Languages meet Prefix Sorting. In Proceedings of the 2020 ACM-SIAM Symposium on Discrete Algorithms, Salt Lake City, UT, USA, 5–8 January 2020; pp. 911–930. [CrossRef]

66. Nerode, A. Linear automaton transformations. *Proc. Am. Math. Soc.* **1958**, *9*, 541–544. [CrossRef]

67. Kosaraju, S.R. Efficient Tree Pattern Matching. In Proceedings of the 30th Annual Symposium on Foundations of Computer Science (SFCS'89), Triangle Park, NC, USA, 30 October–1 November 1989; pp. 178–183.

68. Ferragina, P.; Luccio, F.; Manzini, G.; Muthukrishnan, S. Structuring labeled trees for optimal succinctness, and beyond. In Proceedings of the 46th Annual IEEE Symposium on Foundations of Computer Science (FOCS 2005), Pittsburgh, PA, USA, 23–25 October 2005; pp. 184–196. [CrossRef]

69. Belazzougui, D.; Navarro, G. Optimal Lower and Upper Bounds for Representing Sequences. *ACM Trans. Algorithms* **2015**, *11*, 31. [CrossRef]

70. Arroyuelo, D.; Navarro, G.; Sadakane, K. Stronger Lempel-Ziv Based Compressed Text Indexing. *Algorithmica* **2012**, *62*, 54–101. [CrossRef]

71. Raman, R.; Raman, V.; Rao, S.S. Succinct Indexable Dictionaries with Applications to Encoding K-Ary Trees and Multisets. In Proceedings of the Thirteenth Annual ACM-SIAM Symposium on Discrete Algorithms (SODA'02), San Francisco, CA, USA, 6–8 January 2002.

72. Mantaci, S.; Restivo, A.; Rosone, G.; Sciortino, M. An Extension of the Burrows Wheeler Transform and Applications to Sequence Comparison and Data Compression. In *Combinatorial Pattern Matching*; Apostolico, A., Crochemore, M., Park, K., Eds.; Springer: Berlin/Heidelberg, Germany, 2005; pp. 178–189.

73. Mantaci, S.; Restivo, A.; Rosone, G.; Sciortino, M. An extension of the Burrows—Wheeler transform. *Theor. Comput. Sci.* **2007**, *387*, 298–312. [CrossRef]

74. Mantaci, S.; Restivo, A.; Sciortino, M. An extension of the Burrows Wheeler transform to k words. In Proceedings of the Data Compression Conference, Snowbird, UT, USA, 29–31 March 2005; p. 469.

75. Bowe, A.; Onodera, T.; Sadakane, K.; Shibuya, T. Succinct de Bruijn Graphs. In *Algorithms in Bioinformatics*; Raphael, B., Tang, J., Eds.; Springer: Berlin/Heidelberg, Germany, 2012; pp. 225–235.

76. Sirén, J.; Välimäki, N.; Mäkinen, V. Indexing Graphs for Path Queries with Applications in Genome Research. *IEEE/ACM Trans. Comput. Biol. Bioinform.* **2014**, *11*, 375–388. [CrossRef]

77. Sirén, J. Indexing variation graphs. In Proceedings of the Ninteenth Workshop on Algorithm Engineering and Experiments (ALENEX), Barcelona, Spain, 16–17 January 2017; pp. 13–27.

78. Mäkinen, V.; Cazaux, B.; Equi, M.; Norri, T.; Tomescu, A.I. Linear Time Construction of Indexable Founder Block Graphs. *arXiv* **2020**, arXiv:2005.09342.

79. Na, J.C.; Kim, H.; Min, S.; Park, H.; Lecroq, T.; Léonard, M.; Mouchard, L.; Park, K. FM-index of alignment with gaps. *Theor. Comput. Sci.* **2018**, *710*, 148–157. [CrossRef]

80. Na, J.C.; Kim, H.; Park, H.; Lecroq, T.; Léonard, M.; Mouchard, L.; Park, K. FM-index of alignment: A compressed index for similar strings. *Theor. Comput. Sci.* **2016**, *638*, 159–170. [CrossRef]

81. Durbin, R. Efficient haplotype matching and storage using the positional Burrows—Wheeler transform (PBWT). *Bioinformatics* **2014**, *30*, 1266–1272. [CrossRef] [PubMed]
82. Claude, F.; Navarro, G.; Ordónez, A. The wavelet matrix: An efficient wavelet tree for large alphabets. *Inf. Syst.* **2015**, *47*, 15–32. [CrossRef]
83. Grossi, R.; Gupta, A.; Vitter, J.S. High-Order Entropy-Compressed Text Indexes. In Proceedings of the Fourteenth Annual ACM-SIAM Symposium on Discrete Algorithms (SODA'03), Baltimore, MD, USA, 12–14 January 2003.
84. Baier, U. On Undetected Redundancy in the Burrows-Wheeler Transform. In *Proceedings of the Annual Symposium on Combinatorial Pattern Matching (CPM 2018), Qingdao, China, 2–4 July 2018*; Navarro, G., Sankoff, D., Zhu, B., Eds.; Schloss Dagstuhl–Leibniz-Zentrum fuer Informatik: Dagstuhl, Germany, 2018; Volume 105, pp. 3:1–3:15. [CrossRef]
85. Bentley, J.W.; Gibney, D.; Thankachan, S.V. On the Complexity of BWT-Runs Minimization via Alphabet Reordering. In *Proceedings of the 28th Annual European Symposium on Algorithms (ESA 2020), Pisa, Italy, 7–9 September 2020*; Grandoni, F., Herman, G., Sanders, P., Eds.; Schloss Dagstuhl–Leibniz-Zentrum für Informatik: Dagstuhl, Germany, 2020; Volume 173, pp. 15:1–15:13. [CrossRef]
86. Gibney, D.; Thankachan, S.V. On the Hardness and Inapproximability of Recognizing Wheeler Graphs. In *Proceedings of the 27th Annual European Symposium on Algorithms (ESA 2019), Munich/Garching, Germany, 9–11 September 2019*; Bender, M.A., Svensson, O., Herman, G., Eds.; Schloss Dagstuhl–Leibniz-Zentrum fuer Informatik: Dagstuhl, Germany, 2019; Volume 144, pp. 51:1–51:16. [CrossRef]
87. Gibney, D. Wheeler Graph Recognition on 3-NFAs and 4-NFAs. In Proceedings of the Open Problem Session, International Workshop on Combinatorial Algorithms, Pisa, France, 23–25 July 2020.
88. Alanko, J.; D'Agostino, G.; Policriti, A.; Prezza, N. Wheeler languages. *arXiv* **2020**, arXiv:2002.10303
89. Cotumaccio, N.; Prezza, N. On Indexing and Compressing Finite Automata. In *Proceedings of the 2021 ACM-SIAM Symposium on Discrete Algorithms (SODA'21)*; ACM, Society for Industrial and Applied Mathematics: Philadelphia, PA, USA, 2021; to appear.
90. Rabin, M.O.; Scott, D. Finite Automata and Their Decision Problems. *IBM J. Res. Dev.* **1959**, *3*, 114–125. [CrossRef]

 *algorithms* 

Article

Reversed Lempel–Ziv Factorization with Suffix Trees [†]

Dominik Köppl 

M&D Data Science Center, Tokyo Medical and Dental University, Tokyo 113-8510, Japan;
koeppl.dsc@tmd.ac.jp; Tel.: +81-3-5280-8626
† Parts of this work have been published as part of a Ph.D. Thesis.

Abstract: We present linear-time algorithms computing the reversed Lempel–Ziv factorization [Kolpakov and Kucherov, TCS'09] within the space bounds of two different suffix tree representations. We can adapt these algorithms to compute the longest previous non-overlapping reverse factor table [Crochemore et al., JDA'12] within the same space but pay a multiplicative logarithmic time penalty.

Keywords: longest previous non-overlapping reverse factor table; application of suffix trees; reversed Lempel–Ziv factorization; lossless compression

1. Introduction

The non-overlapping reversed Lempel–Ziv (LZ) factorization was introduced by Kolpakov and Kucherov [1] as a helpful tool for detecting gapped palindromes, i.e., substrings of a given text T of the form $S^R G S$ for two strings S and G, where S^R denotes the reverse of S. This factorization is defined as follows: Given a factorization $T = F_1 \cdots F_z$ for a string T, it is the non-overlapping reversed LZ factorization of T if each factor F_x, for $x \in [1 \mathinner{.\,.} z]$, is either the leftmost occurrence of a character or the longest prefix of $F_x \cdots F_z$ whose reverse has an occurrence in $F_1 \cdots F_{x-1}$. It is a greedy parsing in the sense that it always selects the longest possible such prefix as the candidate for the factor F_x. The factorization can be written like a macro scheme [2], i.e., by a list storing either plain characters or pairs of referred positions and lengths, where a referred position is a previous text position from where the characters of the respective factor can be borrowed. Among all variants of such a left-to-right parsing using the reversed as a reference to the formerly parsed part of the text, the greedy parsing achieves optimality with respect to the number of factors [3] ([Theorem 3.1]) since the reversed occurrence of F_x can be the prefix of any suffix in $F_1 \cdots F_{x-1}$, and thus fulfills the suffix-closed property [3] ([Definition 2.2]).

Kolpakov and Kucherov [1] also gave an algorithm computing the reversed LZ factorization in $\mathcal{O}(n \lg \sigma)$ time using $\mathcal{O}(n \lg n)$ bits of space, by applying Weiner's suffix tree construction algorithm [4] on the reversed text T^R. Later, Sugimoto et al. [5] presented an online factorization algorithm running in $\mathcal{O}(n \lg^2 \sigma)$ time using $\mathcal{O}(n \lg \sigma)$ bits of space. We can also compute the reversed LZ factorization with the longest previous non-overlapping reverse factor table LPnrF storing the longest previous non-overlapping reverse factor for each text position. There are algorithms [6–10] computing LPnrF in linear time for strings whose characters are drawn from alphabets with constant sizes; their used data structures include the suffix automaton [11], the suffix tree of T^R, the position heap [12], and the suffix heap [13]. Finally, Crochemore et al. [14] presented a linear-time algorithm working with integer alphabets by leveraging the suffix array [15]. To find the longest gapped palindromes of the form $S^R G S$ with the length of G restricted in a given interval $\mathcal{I}$, Dumitran et al. [16] ([Theorem 1]) restricted the distance of the previous reverse occurrence relative to the starting position of the respective factor within $\mathcal{I}$ in their modified definition of LPnrF, and achieved the same time and space bounds of [14]. However, all mentioned linear-time approaches use either pointer-based data structures of $\mathcal{O}(n \lg n)$ bits, or multiple integer arrays of length n to compute LPnrF or the reversed LZ factorization.

1.1. Our Contribution

The aim of this paper is to compute the reversed LZ factorization in less space while retaining the linear time bound. For that, we follow the idea of Crochemore et al. [14] ([Section 4]) who built text indexing data structures on $T \cdot \# \cdot T^R$ to compute LPnrF for an artificial character #. However, they need random access to the suffix array, which makes it hard to achieve linear time for working space bounds within $o(n \lg n)$ bits. We can omit the need for random access to the suffix array by a different approach based on suffix tree traversals. As a precursor of this line of research we can include the work of Gusfield [17] ([APL16]) and Nakashima et al. [18]. The former studies the non-overlapping Lempel–Ziv–Storer–Szymanski (LZSS) factorization [2,19] while the latter the Lempel–Ziv-78 factorization [20]. Although their used techniques are similar to ours, they still need $\mathcal{O}(n \lg n)$ bits of space. To actually improve the space bounds, we follow two approaches: On the one hand, we use the leaf-to-root traversals proposed by Fischer et al. [21] ([Section 3]) for the overlapping LZSS factorization [2] during which they mark visited nodes acting as signposts for candidates for previous occurrences of the factors. On the other hand, we use the root-to-leaf traversals proposed in [22] for the leaves corresponding to the text positions of T to find the lowest marked nodes whose paths to the root constitute the lengths of the non-overlapping LZSS factors. Although we mimic two approaches for computing factorizations different to the reversed LZ factorization, we can show that these traversals on the suffix tree of $T \cdot \# \cdot T^R$ help us to detect the factors of the reversed LZ factorization. Our result is as follows:

Theorem 1. *Given a text T of length $n - 1$ whose characters are drawn from an integer alphabet with size $\sigma = n^{\mathcal{O}(1)}$, we can compute its reversed LZ factorization*

- *in $\mathcal{O}(\epsilon^{-1}n)$ time using $(2 + \epsilon)n \lg n + \mathcal{O}(n)$ bits (excluding the read-only text T), or*
- *in $\mathcal{O}(\epsilon^{-1}n)$ time using $\mathcal{O}(\epsilon^{-1}n \lg \sigma)$ bits,*

for a selectable parameter $\epsilon \in (0, 1]$.

On the downside, we have to admit that the results are not based on new tools, but rather a combination of already existing data structures with different algorithmic ideas. On the upside, Theorem 1 presents the first linear-time algorithm computing the reversed LZ factorization using a number of bits linear to the input text T, which is $o(n \lg n)$ bits for $\lg \sigma = o(\lg n)$. Interestingly, this has not yet been achieved for the seemingly easier non-overlapping LZSS factorization, for which we have $\mathcal{O}(\epsilon^{-1}n \log_\sigma^\epsilon n)$ time within the same space bound [22] ([Theorem 1]). We can also adapt the algorithm of Theorem 1 to compute LPnrF, but losing the linear time for the $\mathcal{O}(n \lg \sigma)$-bits solution:

Theorem 2. *Given a text T of length $n - 1$ whose characters are drawn from an integer alphabet with size $\sigma = n^{\mathcal{O}(1)}$, we can compute a 2n-bits representation of its longest previous non-overlapping reverse factor table LPnrF*

- *in $\mathcal{O}(\epsilon^{-1}n)$ time using $(2 + \epsilon)n \lg n + \mathcal{O}(n)$ bits (excluding the read-only text T), or*
- *in $\mathcal{O}(\epsilon^{-1}n \log_\sigma^\epsilon n)$ time using $\mathcal{O}(\epsilon^{-1}n \lg \sigma)$ bits,*

for a selectable parameter $\epsilon \in (0, 1]$. We can augment our LPnrF representation with an $o(n)$-bits data structure to provide constant-time random access to LPnrF entries.

We obtain the $2n$-bits representation of LPnrF with the same compression technique used for the permuted longest common prefix array [23] ([Theorem 1]), see [24] ([Definition 4]) for several other examples.

1.2. Related Work

To put the above theorems into the context of space-efficient factorization algorithms that can also compute factor tables like LPnrF, we briefly list some approaches for different variants of the LZ factorization and of LPnrF. We give Table 1 as an overview. We are aware of approaches to compute the overlapping and non-overlapping LZSS factorization, as well as the longest previous factor (LPF) table LPF [25,26] and the longest previous non-overlapping table LPnF [14]. We can observe in Table 1 that only the overlapping LZSS factorization does not come with a multiplicative $\log_\sigma^\epsilon n$ time penalty when working within $\mathcal{O}(\epsilon^{-1} n \lg \sigma)$ bits. Note that the time and space bounds have an additional multiplicative ϵ^{-1} penalty (unlike described in the references therein) because the currently best construction algorithms of the compressed suffix tree (described later in Section 2) works in $\mathcal{O}(\epsilon^{-1} n)$ time and needs $\mathcal{O}(\epsilon^{-1} n \lg \sigma)$ bits of space [27] ([Section 6.1]).

Regarding space-efficient algorithms computing the LZSS factorization, we are aware of the linear-time algorithm of Goto and Bannai [28] using $n \lg n + \mathcal{O}(\sigma \lg n)$ bits of working space. For ϵn bits of space, Kärkkäinen et al. [29] can compute the factorization in $\mathcal{O}(n \lg n \lg \lg \sigma)$ time, which got improved to $\mathcal{O}(n(\lg \sigma + \lg \lg n))$ by Kosolobov [30]. Finally, the algorithm of Belazzougui and Puglisi [31] uses $\mathcal{O}(n \lg \sigma)$ bits of working space and $\mathcal{O}(n \lg \lg \sigma)$ time.

Another line of research is the online computation of LPF. Here, Okanohara and Sadakane [32] gave a solution that works in $n \lg \sigma + \mathcal{O}(n)$ bits of space and needs $\mathcal{O}(n \lg^3 n)$ time. This time bound got recently improved to $\mathcal{O}(n \lg^2 n)$ by Prezza and Rosone [33].

Table 1. Complexity bounds of related approaches described in Section 1.2 for a selectable parameter $\epsilon \in (0, 1]$.

$(1+\epsilon)n \lg n + \mathcal{O}(n)$ Bits of Working Space (Excluding the Read-Only Text T)		
Reference	**Type**	**Time**
[21] ([Corollary 3.7])	overlapping LZSS	$\mathcal{O}(\epsilon^{-1} n)$
[34] ([Lemma 6])	LPF	$\mathcal{O}(\epsilon^{-1} n)$
[22] ([Theorem 1])	non-overlapping LZSS	$\mathcal{O}(\epsilon^{-1} n)$
[22]([Theorem 3])	LPnF	$\mathcal{O}(\epsilon^{-1} n)$
$\mathcal{O}(\epsilon^{-1} n \lg \sigma)$ **Bits of Working Space**		
Reference	**Type**	**Time**
[21] ([Corollary 3.4])	overlapping LZSS	$\mathcal{O}(\epsilon^{-1} n)$
[34] ([Lemma 6])	LPF	$\mathcal{O}(\epsilon^{-1} n \log_\sigma^\epsilon n)$
[22] ([Theorem 1])	non-overlapping LZSS	$\mathcal{O}(\epsilon^{-1} n \log_\sigma^\epsilon n)$
[22] ([Theorem 3])	LPnF	$\mathcal{O}(\epsilon^{-1} n \log_\sigma^\epsilon n)$

1.3. Structure of this Article

This article is structured as follows: In Section 2, we start with the introduction of the suffix tree representations we build on the string $T \cdot \# \cdot T^R$, and introduce the reversed LZ factorization in Section 3. We present in Section 3.2 our solution for the claim of Theorem 1 without the referred positions, which we compute subsequently in Section 3.3. Finally, we introduce LPnrF in Section 4, and give two solutions for Theorem 2. One is a derivation of our reversed-LZ factorization algorithm of Section 3.2.2 (cf. Section 4.1), the other is a translation of [14] ([Algorithm 2]) to suffix trees (cf. Section 4.2).

2. Preliminaries

With lg we denote the logarithm $\log_2$ to base two. Our computational model is the word RAM model with machine word size $\Omega(\lg n)$ for a given input size n. Accessing a word costs $\mathcal{O}(1)$ time.

Let T be a text of length $n - 1$ whose characters are drawn from an integer alphabet $\Sigma = [1 .. \sigma]$ with $\sigma = n^{\mathcal{O}(1)}$. Given $X, Y, Z \in \Sigma^*$ with $T = XYZ$, then X, Y and Z are called a prefix, substring and suffix of T, respectively. We call $T[i ..]$ the i-th suffix of T, and denote a substring $T[i]T[i+1] \cdots T[j]$ with $T[i .. j]$. For $i > j$, $[i .. j]$ and $T[i .. j]$ denote the empty set and the empty string, respectively. The reverse T^R of T is the concatenation $T^R := T[n-1]T[n-2] \cdots T[1]$. We further write $T[i .. j]^R := T[j]T[j-1] \cdots T[i]$.

Given a character $c \in \Sigma$, and an integer j, the rank query $T.\mathrm{rank}_c(j)$ counts the occurrences of c in $T[1..j]$, and the select query $T.\mathrm{select}_c(j)$ gives the position of the j-th c in T, if it exists. We stipulate that $\mathrm{rank}_c(0) = \mathrm{select}_c(0) = 0$. If the alphabet is binary, i.e., when T is a bit vector, there are data structures [35,36] that use $o(|T|)$ extra bits of space, and can compute rank and select in constant time, respectively. There are representations [37] with the same constant-time bounds that can be constructed in time linear in $|T|$. We say that a bit vector has a rank-support and a select-support if it is endowed by data structures providing constant time access to rank and select, respectively.

From now on, we assume that there exist two special characters # and \$ that do not appear in T, with $\$ < \# < c$ for every character $c \in \Sigma$. Under this assumption, none of the suffixes of $T \cdot \#$ and $T^R \cdot \$$ has another suffix as a prefix. Let $R := T \cdot \# \cdot T^R \cdot \$$. R has length $|R| = 2|T| + 2 = 2n$.

The suffix tree ST of R is the tree obtained by compacting the suffix trie, which is the trie of all suffixes of R. ST has $2n$ leaves and at most $2n - 1$ internal nodes. The string stored in a suffix tree edge e is called the label of e. The children of a node v are sorted lexicographically with respect to the labels of the edges connecting the children with v. We identify each node of the suffix tree by its pre-order number. We do so implicitly such that we can say, for instance, that a node v is marked in a bit vector B, i.e., $B[v] = 1$, but actually have $B[i] = 1$, where i is the pre-order number of v. The string label of a node v is defined as the concatenation of all edge labels on the path from the root to v; v's string depth, denoted by $\mathrm{str\_depth}(v)$, is the length of v's string label. The operation $\mathrm{suffixlink}(v)$ returns the node with string label $S[2..]$ or the root node, given that the string label of v is S with $|S| \geq 2$ or a single character, respectively. suffixlink is undefined for the root node.

The leaf corresponding to the i-th suffix $R[i..]$ is labeled with the suffix number $i \in [1..2n]$. We write $\mathrm{sufnum}(\lambda)$ for the suffix number of a leaf λ. The leaf-rank is the preorder rank ($\in [1..2n]$) of a leaf among the set of all ST leaves. For instance, the leftmost leaf in ST has leaf-rank 1, while the rightmost leaf has leaf-rank $2n$. To avoid confusing the leaf-rank with the suffix number of a leaf, let us bear in mind that the leaf-ranks correspond to the lexicographical order of the suffixes (represented by the leaves) in R, while the suffix numbers induce a ranking based on the text order of R's suffixes. In this context, the function $\mathrm{suffixlink}(\lambda)$ returns the leaf whose suffix number is $\mathrm{sufnum}(\lambda) + 1$. The reverse function of suffixlink on leaves is $\mathrm{prev\_leaf}(\lambda)$ that returns the leaf whose suffix number is $\mathrm{sufnum}(\lambda) - 1$, or $2n$ if $\mathrm{sufnum}(\lambda) = 1$ (We do not need to compute $\mathrm{suffixlink}(\lambda)$ for a leaf with $\mathrm{sufnum}(\lambda) = 2n$, but want to compute $\mathrm{prev\_leaf}(\lambda)$ for the border case $\mathrm{sufnum}(\lambda) = 1$.).

In this article, we focus on the following two ST representations: the compressed suffix tree (CST) [23,38] and the succinct suffix tree (SST) [21] ([Section 2.2.3]). Both can be computed in $\mathcal{O}(\epsilon^{-1}n)$ time, where the former is due to a construction algorithm given by Belazzougui et al. [27] ([Section 6.1]), and the latter due to [21] ([Theorem 2.8]), see Table 2. These two representations provide some of the above described operations in the time bounds listed in Table 3. Each representation additionally stores the pointer smallest_leaf to the leaf with suffix number 1, and supports the following operations in constant time, independent of ϵ:

leaf_rank(λ) returns the leaf-rank of the leaf λ;

depth(v) returns the depth of the node v, which is the number of nodes on the path between v and the root (exclusive) such that root has depth zero;

level_anc(λ, d) returns the level-ancestor of the λ on depth d; and

lca(u, v) returns the lowest common ancestor (LCA) of u and v.

As previously stated, we implicitly represent nodes by their pre-order numbers such that the above operations actually take pre-order numbers as arguments.

Table 2. Construction time and needed space in bits for the succinct suffix tree (SST) and compressed suffix tree (CST) representations, cf. [21] ([Section 2.2]).

	SST	CST
Time	$\mathcal{O}(n/\epsilon)$	$\mathcal{O}(\epsilon^{-1}n)$
Space	$(2+\epsilon)n\lg n + \mathcal{O}(n)$	$\mathcal{O}(\epsilon^{-1}n\lg\sigma)$

Table 3. Time bounds for certain operations needed by our LZ factorization algorithms. Although not explicitly mentioned in [21], the time for prev_leaf is obtained with the Burrows–Wheeler transform [39] stored in the CST [38] ([A.1]) by constant-time partial rank queries, see [27] ([Section 3.4]) or [38] ([A.4]).

Operation	SST Time	CST Time
sufnum(λ)	$\mathcal{O}(1/\epsilon)$	$\mathcal{O}(n)$
str_depth(v)	$\mathcal{O}(1/\epsilon)$	$\mathcal{O}(\text{str\_depth}(v))$
suffixlink(v)	$\mathcal{O}(1/\epsilon)$	$\mathcal{O}(1)$
prev_leaf	$\mathcal{O}(1/\epsilon)$	$\mathcal{O}(1)$

3. Reversed LZ Factorization

A factorization of T of size z partitions T into z substrings $F_1 \cdots F_z = T$. Each such substring F_x is called a factor. A factorization is called reversed LZ factorization if each factor F_x is either the leftmost occurrence of a character or the longest prefix of $F_x \cdots F_z$ that occurs at least once in $(F_1 \cdots F_{x-1})^R$, for $x \in [1 \mathinner{.\,.} z]$. A similar but much well-studied factorization is the non-overlapping LZSS factorization, where each factor F_x is either the leftmost occurrence of a character or the longest prefix of $F_x \cdots F_z$ that occurs at least once in $F_1 \cdots F_{x-1}$, for $x \in [1 \mathinner{.\,.} z]$. See Figure 1 for an example and a comparison of both factorizations. In what follows, let z denote the number of reversed-LZ factors of T.

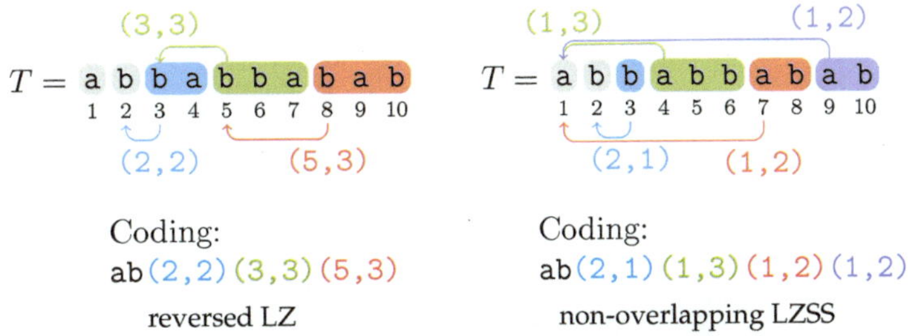

Figure 1. The reversed LZ and the non-overlapping LZSS factorization of the string $T = \texttt{abbabbabab}$. A factor F is visualized by a rounded rectangle. Its coding consists of a mere character if it has no reference; otherwise, its coding consists of its referred position p and its length ℓ such that $F = T[p-\ell+1 \mathinner{.\,.} p]^R$ for the reversed LZ factorization, and $F = T[p \mathinner{.\,.} p+\ell-1]$ for the non-overlapping LZSS factorization.

3.1. Coding

We classify factors into fresh and referencing factors: We say that a factor is fresh if it is the leftmost occurrence of a character. We call all other factors referencing. A referencing factor F_x has a reference pointing to the ending position of its longest previous non-overlapping reverse occurrence; as a tie break, we always select the leftmost such ending position. We call this ending position the referred position of F_x. More precisely, the referred position of a factor $F_x = T[i \mathinner{.\,.} i+\ell-1]$ is the smallest text position j with $j \le i-1$ and $T[j-\ell+1 \mathinner{.\,.} j]^R = T[i \mathinner{.\,.} i+\ell-1]$. If we represent each referencing factor as a pair consisting of its referred position and its length, we obtain the coding shown in Figure 1.

Although our tie breaking rule selecting the leftmost position among all candidates for the referred position seems up to now arbitrary, it technically simplifies the algorithm in that we only have to index the very first occurrence.

3.2. Factorization Algorithm

In the following, we describe our factorization algorithm working with ST. This algorithm performs traversals on paths connecting leaves with the root, during which it marks certain nodes. One kind of these marked nodes are phrase leaves: A phrase leaf is a leaf whose suffix number is the starting position of a factor. We say that a phrase leaf λ corresponds to a factor F if the suffix number of ℓ is the starting position of F. We call all other leaves non-phrase leaves. Another kind are witnesses, a notion borrowed from [21] ([Section 3]): Witnesses are nodes that create a connection between referencing factors and their referred positions. We formally define them as follows: given λ is the phrase leaf corresponding to a referencing factor F, the witness w of F is the LCA of λ and a leaf with suffix number $2n - j$ (with $j \in [1 .. n - 1]$) such that $T[j - \text{str\_depth}(w) + 1 .. j]^R$ is the longest substring in $T[1 .. \text{sufnum}(\lambda) - 1]^R$ that is a prefix of $T[\text{sufnum}(\lambda) ..]$. The smallest such j is the referred position of λ, which is needed for the coding in Section 3.1. See Figure 2 for a sketch of the setting. In what follows, we show that the witness of a referencing factor F is the node whose string label is F. Generally speaking, for each substring S of T, there is always a node whose string label has S as a prefix, but there maybe no node whose string label is precisely S. This is in particular the case for the non-overlapping LZSS factorization [22] ([Section 3.1]). Here, we can make use of the fact that the suffix number $2n - j$ for a referred position j is always larger than the length of T, which we want to factorize:

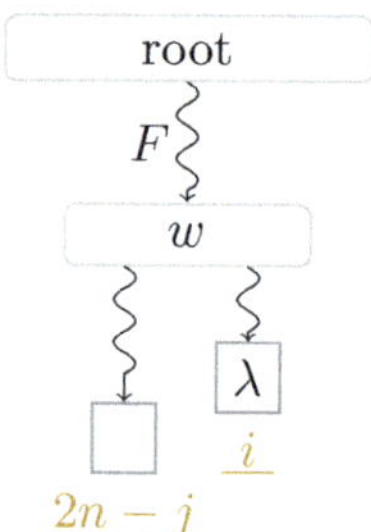

Figure 2. Witness node w of a referencing factor F starting at text position i. Given j is the referred position of F, the witness w of F is the node in the suffix tree having (a) F as a prefix of its string label and (b) the leaves with suffix numbers $2n - j$ and i in its subtree. Lemma 1 shows that w is uniquely defined to be the node whose string label is F.

Lemma 1. *The witness of each referencing factor exists and is well-defined.*

Proof. To show that each referencing factor is indeed the string label of an ST node, we review the definition of right-maximal repeats: A right-maximal repeat is a substring of R having at least two occurrences $R[i_1 .. i_1 + \ell - 1]$ and $R[i_2 .. i_2 + \ell - 1]$ with $R[i_1 + \ell] \neq R[i_2 + \ell]$. A right-maximal repeat is the string label of an ST node since this node has at least two children; those two children are connected by edges whose labels start with $R[i_1 + \ell]$ and $R[i_2 + \ell]$, respectively. It is therefore sufficient to show that each factor F is a right-maximal repeat. Given j is the referred position of $F = T[i .. i + |F| - 1]$, $F = T[j - |F| + 1 .. j]^R = R[2n - j .. 2n - j + |F| - 1]$. If $j = |F|$, then $T[i + |F|] \neq R[2n - j + |F|] = \$$, and thus F is a right-maximal repeat. For the other case that $j \geq |F| + 1$, assume that F is not a right-maximal repeat. Then $T[i + |F|] = R[2n - j + |F|] = T[j - |F|]$. However, this means that F is not the longest reversed factor being a prefix of $T[i ..]$, a contradiction. We visualized the situation in Figure 3. $\square$

Consequently, the referred position of a factor $F_x = T[i..i+\ell-1]$ is the smallest text position j in T with $j \leq i-1$ and one of the two equivalent conditions hold:

- $T[j-\ell+1..j]^{\mathrm{R}} = T[i..i+\ell-1]$; or
- $R[i..]$ and $R[2n-j..]$ have the longest common prefix of length ℓ.

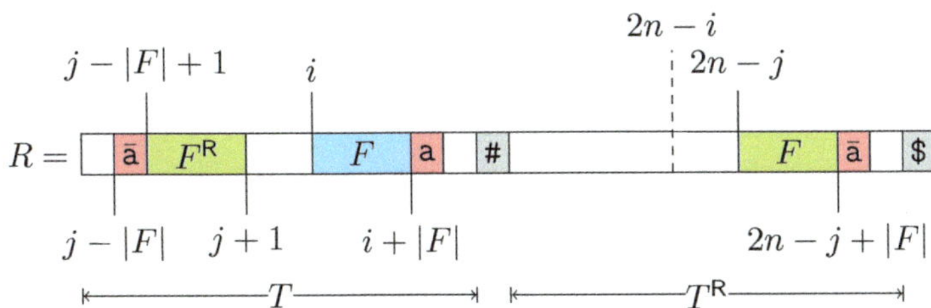

Figure 3. A reversed-LZ factor F starting at position i in R with a referred position $j \geq |F|+1$. If $a = \bar{a}$ with $a, \bar{a} \in \Sigma$, then we could extend F by one character, contradicting its definition to be the longest prefix of $T[i..]$ whose reverse occurs in $T[1..i-1]$. Hence, $a \neq \bar{a}$ and F is a right-maximal repeat.

3.2.1. Overview

We explain our factorization algorithm in terms of a cooperative game with two players (We use this notation only for didactic purposes; the terminology must not be confused with game theory. Here, the notion of player is basically a subroutine of the algorithm having private and shared variables.), whose pseudo code we sketched in Algorithm 1. Player 1 and Player 2 are allowed to access the leaves with suffix numbers in the ranges $[1..n]$ and $[n..2n-1]$, respectively. Player 1 (resp. Player 2) starts at the leaf with the smallest (resp. largest) suffix number, and is allowed to access the leaf with the subsequently next (resp. previous) suffix number via suffixlink (resp. prev_leaf). Hence, Player 1 simulates a linear forward scan in the text T, while Player 2 simulates a linear backward scan in T^{R}. Both players take turns at accessing leaves at the same pace. To be more precise, in the i-th turn, Player 1 processes the leaf with suffix number i, whereas Player 2 processes the leaf with suffix number $2n-i$. In one turn, a player accesses a leaf λ and maybe performs a traversal on the path connecting the root with λ. For such a traversal, we use level ancestor queries to traverse each node on the path in constant time. Whenever Player 2 accesses the leaf with suffix number n (shared among both players), the game ends; at that time both players access the same leaf (cf. Line 6 in Algorithm 1). In the following, we call this game a pass (with the meaning that we pass all relevant text positions). Depending on the allowed working space, our algorithm consists of one or two passes (cf. Section 3.3). The goal of Player 2 is to keep track of all nodes she visits. Player 2 does this by maintaining a bit vector B_{V} of length $4n$ such that $B_{\mathrm{V}}[v]$ stores whether a node v has already been visited by Player 2, where we represent a node v by its pre-order number when using it as an index of a bit vector. To keep things simple, we initially mark the root node in B_{V} at the start of each pass. By doing so, after the i-th turn of Player 2 we can read any substring of $T[1..i]^{\mathrm{R}}$ by a top-down traversal from the suffix tree root, only visiting nodes marked in B_{V}. This is because of the invariant that the set of nodes marked in B_{V} is upper-closed, i.e., if a node v is marked in B_{V}, then all its ancestors are marked in B_{V} as well.

The goal of Player 1 is to find the phrase leaves and the witnesses. For that, she maintains two bit vectors B_{L} and B_{W} of length n and $4n$, respectively, whose entries are marked similarly to B_{V} by using the suffix numbers ($\in [1..n]$) of the leaves accessed by Player 1 and preorder numbers of the internal nodes. We initially mark smallest_leaf in B_{L} since text position 1 is always the starting position of the fresh factor F_1. By doing so, after the i-th turn of Player 1 we know the ending positions of those factors contained in $T[1..i]$, which are marked in B_{L}. To sum up, after the i-th turn of both players we know the computed factors starting at text positions up to i thanks to Player 1, and can find the factor lengths thanks to Player 2, which we explain in detail in Section 3.2.2. There, we will show that the actions of Player 2 allow Player 1 to determine the starting position of the next factor. For that, she computes the string depth of the lowest ancestor marked in B_V of the previously visited phrase leaf. See Appendix A.

As a side note: since we are only interested in the factorization of $T[1 .. n - 1]$ (omitting the appended # at position n), we do not need Player 1 to declare the leaf with suffix number n a phrase leaf. We also terminate the algorithm when both players meet at position n without checking whether we have found a new factor starting at position n.

Algorithm 1: Algorithm of Section 3.2.2 computing the non-overlapping reversed LZ factorization. The function max_sufnum is described in Section 3.3.

```
1  ST ← suffix tree of R = T · # · T^R · $
2  λ^R ← prev_leaf(prev_leaf(smallest_leaf))              ▷ sufnum(λ^R) = 2n − 1
3  λ ← smallest_leaf
4  B_V[root node] ← 1        ▷ at the beginning, only the root node is marked in B_V
5  B_L[1] ← 1                                      ▷ |F_1| starts at text position 1
6  while λ ≠ λ^R do                              ▷ we stop after having parsed T[1 .. n]
7      if B_L[sufnum(λ)] = 1 then                              ▷ turn of Player 1
8          d ← 0
9          while B_V[level_anc(λ, d + 1)] = 1 do d ← d + 1
10         w ← level_anc(λ, d)                    ▷ w is lowest node marked in B_V
11         if w is the root then
12             output fresh factor
13             B_L[sufnum(λ) + 1] ← 1       ▷ next factor starts directly after sufnum(λ)
14         else                                         ▷ w is the witness of λ
15             output length str_depth(w)
16             B_L[sufnum(λ) + str_depth(w)] ← 1
17             output referred position 2n − max_sufnum(w)         ▷ for the one-pass
                   variant
18             B_W[w] ← 1                                  ▷ for the two-pass variant
19     λ ← suffixlink(λ)                                ▷ end of Player 1's turn
20     foreach node v on the path from λ^R up to the root do          ▷ turn of Player 2
21         if B_V[v] = 1 then break     ▷ end turn on reaching an already marked node
22         B_V[v] ← 1
23     λ^R ← prev_leaf(λ^R)                               ▷ end of Player 2's turn
```

3.2.2. One-Pass Algorithm in Detail

In detail, a pass works as follows: at the start, Player 1 and Player 2 select smallest_leaf and prev_leaf(prev_leaf(smallest_leaf)), i.e., the leaves with suffix numbers 1 and $2n - 1$, respectively. Now the players take action in alternating turns, starting with Player 1. Nevertheless, we first explain the actions of Player 2, since Player 2 acts independently of Player 1, while Player 1's actions depend on Player 2.

Suppose that Player 2 is at a leaf λ^R (cf. Line 20 of Algorithm 1). Player 2 traverses the path from λ^R to the root upwards and marks all visited nodes in B_V until arriving at a node v already marked in B_V (such a node exists since we mark the root in B_V at the beginning of a pass.). When reaching the marked node v, we end the turn of Player 2, and move Player 2 to prev_leaf(λ^R) at Line 23 (and terminate the whole pass in Line 6 when this leaf has suffix number n). The `foreach` loop (Line 20) of the algorithm can be more verbosely expressed with a loop iterating over all depth offsets d in increasing order while computing $v \leftarrow$ level_anc(λ^R, d) until either reaching the root or a node marked in B_V. Subsequently, the turn of Player 1 starts (cf. Line 7). We depict the state after the first turn of Player 2 in Figure 4.

If Player 1 is at a non-phrase leaf λ, we skip the turn of Player 1, move Player 1 to suffixlink(λ) at Line 19, and let Player 2 take action. Now suppose that Player 1 is at a phrase leaf λ corresponding to a factor F. Then we traverse the path from the root to λ downwards to find the lowest ancestor w of λ marked in B_V. If w is the root node, then F is a fresh factor (cf. Line 11), and we know that the next factor starts immediately after F (cf. Line 13). Consequently, the leaf suffixlink(λ) is a phrase leaf. Otherwise, w is the witness of λ, and str_depth(w) = $|F|$ (cf. Line 14). Hence, sufnum(λ) + str_depth(w) is the suffix number of the phrase leaf $\tilde{\lambda}$ that Player 1 will subsequently access. We therefore mark w and sufnum($\tilde{\lambda}$) = sufnum(λ) + str_depth(w) in B_W and in B_L, respectively (cf. Lines 16 and 18). We depict the fifth turn of our running example in Figure 5, during which Player 1 marks a witness node. Finally, we end the turn of Player 1, move Player 1 to suffixlink(λ) at Line 19, and let Player 2 take action.

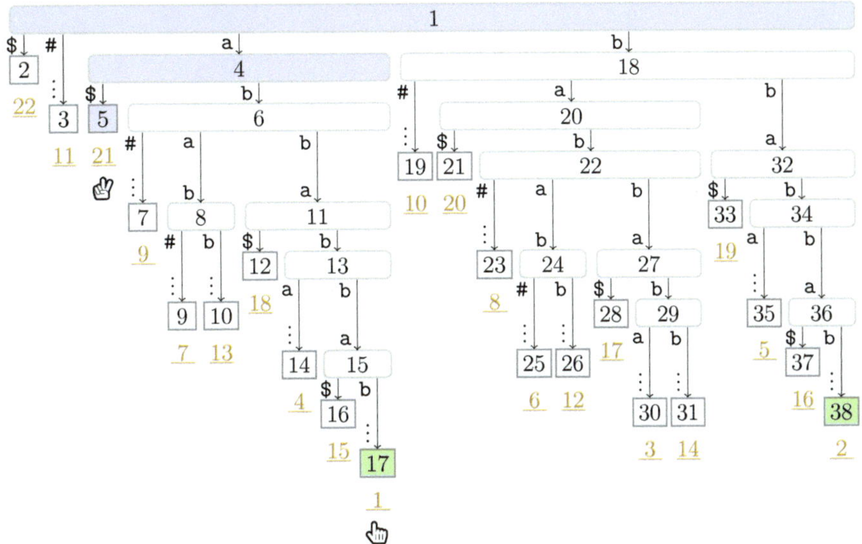

Figure 4. Suffix tree of $T\# \cdot T^R \cdot \$$ used in Section 3.2, where $T =$ abbabbabab is our running example. The nodes are labeled by their preorder numbers. The suffix number of each leaf λ is the underlined number drawn in dark yellow below λ. We trimmed the label of each edge to a leaf having more than two characters and display only the first character and the vertical dots ':' as a sign of omission. The tree shows the state of Algorithm 1 after the first turn of both players. The nodes visited by Player 2 are colored in blue (⬜), the phrase leaves are colored in green (🟩). Player 1 and 2 are represented by the hands 👆 and 👇, respectively, pointing to the respective leaves they visited during the first turn.

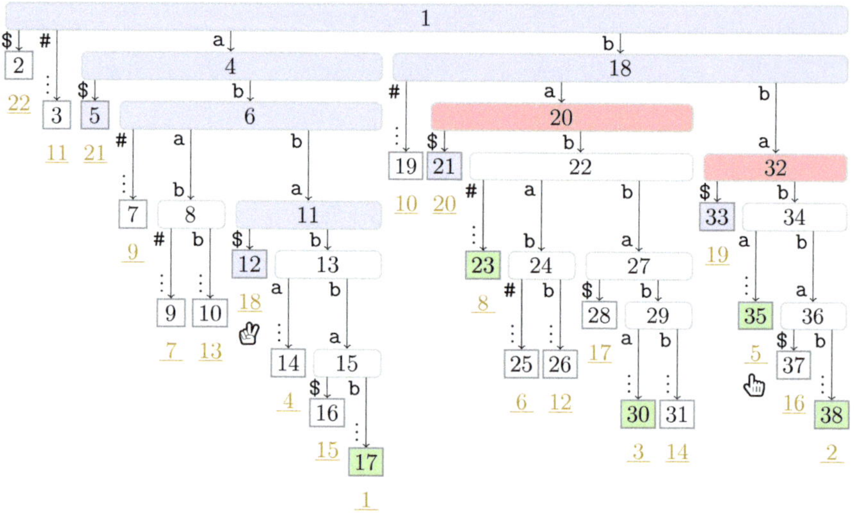

Figure 5. Continuation of Figure 4 with the state at the fifth turn of Player 1. Additionally to the coloring used in Figure 4, witnesses are colored in red (▭). In this figure, Player 1 just finished her turn on making the node with preorder number 32 the witness w of the leaf with suffix number 5. With w we know that the factor starting at text position 5 has the length str_depth(w) and that the next phrase leaf has suffix number 8. For visualization purposes, we left the hand (☝) of Player 2 below the leaf of her last turn.

Correctness. When Player 1 accesses the leaf λ with suffix number i, Player 2 has processed all leaves with suffix numbers $[2n - i + 1 .. 2n - 1]$. Due to the leaf-to-root traversals of Player 2, each node marked in B_V has a leaf with a suffix number in $[2n - i + 1 .. 2n - 1]$ in its subtree. In particular, a node w is marked in B_V if and only if the string label of w is a substring of $R[2n - i + 1 .. 2n - 1]$. Because $R[2n - i + 1 .. 2n - 1]^R = T[1 .. i - 1]$, the longest prefix of $T[i ..]$ having a reversed occurrence in $T[1 .. i - 1]$ is therefore one of the string labels of the nodes marked in B_V. In particular, we search the longest string label among those nodes, which we obtain with the lowest ancestor of λ marked in B_V.

3.2.3. Time Complexity

First, let us agree on that we never compute the suffix number of a leaf since this is a costly operation for CST (cf. Table 3). Although we need the suffix numbers at multiple occasions, we can infer them if each player maintains a counter for the suffix number of the currently visited leaf. A counter is initialized with 1 (resp. $2n - 1$) and becomes incremented (resp. decremented) by one when moving to the succeeding (resp. preceding) leaf in suffix number order. This works since both players traverse the leaves linearly in the order of the suffix numbers (either in ascending or descending order).

Player 2 visits n leaves, and visits only unvisited nodes during a leaf-to-root traversal. Hence, Player 2's actions take $\mathcal{O}(n)$ overall time.

Player 1 also visits n leaves. Since Player 1 has no business with the non-phrase leaves, we only need to analyze the time spent by Player 1 for a phrase leaf corresponding to a factor F: If F is fresh, then the root-to-leaf traversal ends prematurely at the root, and hence we can determine in constant time whether F is fresh or not. If F is referencing, we descend from the root to the lowest ancestor w marked in B_V, and compute str_depth(w) to determine the suffix number of the next phrase leaf (cf. Line 15 of Algorithm 1). Since depth(w) $\leq$ str_depth(w), we visit at most $|F| + 1$ nodes before reaching w. Computing str_depth(w) takes $\mathcal{O}(1/\epsilon)$ time for the SST, and $\mathcal{O}(|F|)$ time for the CST. This seems costly, but we compute str_depth(w) for each factor only once. Since the sum of all factor lengths is n, we spend $\mathcal{O}(n + z/\epsilon)$ time or $\mathcal{O}(n)$ time for computing all factor lengths when using the SST or the CST, respectively. We finally obtain the time bounds stated in Theorem 1 for computing the factorization.

3.3. Determining the Referred Position

Up to now, we can determine the reversed-LZ factors $F_1 \cdots F_z = T$ with B_L marking the starting position of each factor with a one. Yet, we have not the referred positions necessary for the coding of the factors (cf. Section 3.1). To obtain them, we have two options: The first option is easier but comes with the requirement for a support data structure on ST for the operation

max_sufnum(v) returning the maximum among all suffix numbers of the leaves in the subtree rooted in v.

We can build such a support data structure in $\mathcal{O}(\epsilon^{-1}n)$ time (resp. $\mathcal{O}(\epsilon^{-1}n \log_\sigma^\epsilon n)$ time) using $\mathcal{O}(n)$ bits to support max_sufnum in $\mathcal{O}(\epsilon^{-1})$ time (resp. $\mathcal{O}(\epsilon^{-1} \log_\sigma^\epsilon n)$ time) for the SST (resp. CST); see [22] ([Section 3.3]). Being able to query max_sufnum, we can directly compute the referred position of a factor F when discovering its witness w during a turn of Player 1 by max_sufnum(w). max_sufnum(w) gives us the suffix number of a leaf that has already been accessed by Player 2 since Player 2 accesses the leaves in descending order with respect to the suffix numbers, and w must have already been accessed by Player 2 during a leaf-to-root traversal (otherwise w would not have been marked in B_V). Since $R[\text{max\_sufnum}(w) \mathinner{\ldotp\ldotp} \text{max\_sufnum}(w) + \text{str\_depth}(w) - 1] = F^R$, the referred position of F is $2n - \text{max\_sufnum}(w)$. Consequently, we can compute the coding of the factors during a single pass (cf. Line 17 of Algorithm 1), and are done when the pass finishes.

The second option does not need to compute max_sufnum and retains the linear time bound when using CST. Here, the idea is to run an additional pass, whose pseudo code is given in Algorithm 2. For this additional pass, we do the following preparations: Let z_W be the number of witnesses, which is at most z since there can be multiple factors having the same witness. We keep B_L and B_W marking the phrase leaves and the witnesses, respectively. However, we clear B_V such that Player 2 has again the job to log her visited nodes in B_V. We augment B_W with a rank-support such that we can enumerate the witnesses with ranks from 1 to at most z_W, which we call the witness rank. We additionally create an array W of $z_W \lg n$ bits. We want $W[B_W.\text{rank}_1(w)]$ to store the referred position $2n - \text{max\_sufnum}(w) \in [1 \mathinner{\ldotp\ldotp} n - 1]$ for each witness w such that we can read the respective referred position from W when Player 1 accesses w. We assign the task for maintaining W to Player 2. Player 2 can handle this task by taking additional action when visiting a witness (i.e., a node marked in B_W) during a leaf-to-root traversal: When visiting a witness node w with witness rank i from a leaf λ, we write $W[i] \leftarrow 2n - \text{sufnum}(\lambda)$ if w is not yet marked in B_V (cf. Line 15 in Algorithm 2). Like before, Player 2 terminates her turn whenever she visits an already visited node. The actions of Player 1 differ in that she no longer needs to compute B_L and B_W: When Player 1 visits a phrase leaf λ, she locates the lowest ancestor w of λ marked in B_V, which has to be marked in B_W, too (as a side note: storing the depth of the witness of each phrase leaf in a list, sorted by the suffix numbers of these leaves, helps us to directly jump to the respective witness in constant time. We can encode this list as a bit vector of length $\mathcal{O}(n)$ by storing each depth in unary coding (cf. [22] ([Section 3.4])). Nevertheless, we can afford the root-to-witness traversals of Player 1 since we visit at most $\sum_{x=1}^{z} |F_x| = n$ nodes in total.). With the rank-support on B_W, we can compute w's witness rank i, and obtain the referred position of λ with $W[i]$ (cf. Line 10 of Algorithm 2). We show the final state after the first pass in Figure 6, together with W computed in the second pass.

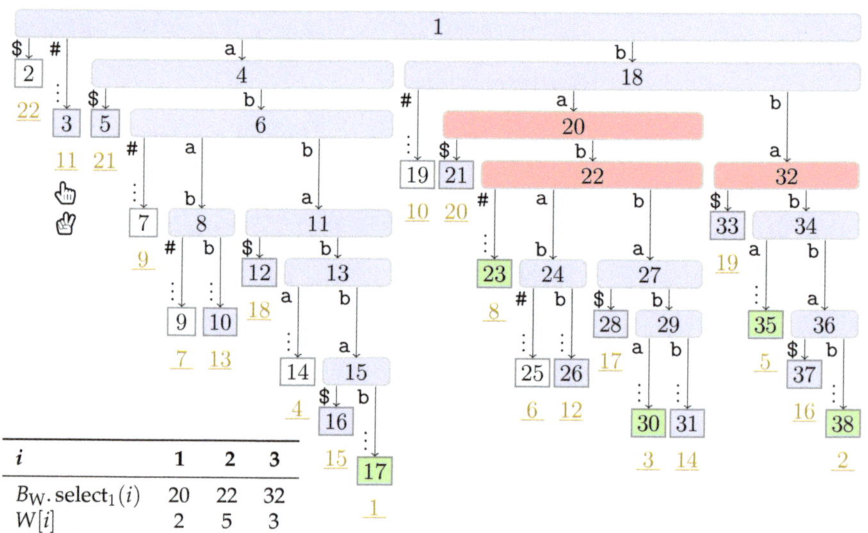

i	1	2	3
$B_W.\text{select}_1(i)$	20	22	32
$W[i]$	2	5	3

Figure 6. State of our running example at termination of Algorithm 1. We have computed the bit vector B_L of length $n = 11$ storing a one at the entries $1, 2, 3, 5$, and 8, i.e., the suffix numbers of the phrase leaves, which are marked in green (□), and the bit vector B_W of length 38 (the maximum preorder number of an ST node) storing a one at the entries $20, 22$, and 32, i.e., the preorder numbers of the witnesses, which are colored red (□). During the second pass described in Section 3.3, we compute W storing the referred positions in the order of the witness ranks (left table).

Overall, the time complexity is $\mathcal{O}(\epsilon^{-1}n)$ time when working with either the SST or the CST. We use $o(n)$ additional bits of space for the rank-support of B_W, but costly $z_W \lg n$ bits for the array W. However, we can bound z_W by $\mathcal{O}(n \lg \sigma / \lg n)$ since z_W is the number of distinct reversed LZ factors, and by an enumeration argument [40] ([Thm. 2]), a text of length n can be partitioned into at most $\mathcal{O}(n / \log_\sigma n)$ distinct factors. Hence, we can store W in $z_W \lg n = \mathcal{O}(n \lg \sigma)$ bits of space. With that, we finally obtain the working space bound of $\mathcal{O}(\epsilon^{-1}n \lg \sigma)$ bits for the CST solution as claimed in Theorem 1.

4. Computing LPnrF

The longest previous non-overlapping reverse factor table $\text{LPnrF}[1 \mathinner{.\,.} n]$ is an array such that $\text{LPnrF}[i]$ is the length of the longest prefix of $T[i \mathinner{.\,.}] \cdot \#$ occurring as a substring of $T[1 \mathinner{.\,.} i-1]^R$. (Appending $\#$ at the end is not needed, but simplifies the analysis for $T[1 \mathinner{.\,.} n-1] \cdot \#$ having precisely n characters.) Having LPnrF, we can iteratively compute the reversed LZ factorization because $F_x = T[k_x \mathinner{.\,.} k_x + \max(0, \text{LPnrF}[k_x] - 1)]$ with $k_x := 1 + \sum_{y=1}^{x-1} |F_y|$ for $x \in [1 \mathinner{.\,.} z]$.

The counterpart of LPnrF for the non-overlapping LZSS factorization is the longest previous non-overlapping factor table $\text{LPnF}[1 \mathinner{.\,.} n]$, which is defined similarly, but stores the maximal length of the longest common prefix (LCP) of $T[i \mathinner{.\,.}]$ with all substrings $T[j \mathinner{.\,.} i-1]$ for $j \in [1 \mathinner{.\,.} i-1]$. See Table 4 for a comparison. Analogously to [34] ([Corollary 5]) or [24] ([Definition 4]) for the longest previous factor table LPF [22,26] ([Lemma 1]) for LPnF, LPnrF holds the following property:

Lemma 2 ([14] (Lemma 2)). $\text{LPnrF}[i-1] - 1 \leq \text{LPnrF}[i] \leq n - i$ *for* $i \in [2 \mathinner{.\,.} n]$.

Hence, we can encode LPnrF in $2n$ bits by writing the differences $\text{LPnrF}[i] - \text{LPnrF}[i-1] + 1 \geq 0$ in unary, obtaining a bit sequence of (a) n ones for the n entries and (b) $\sum_{i=2}^{n}(\text{LPnrF}[i] - \text{LPnrF}[i-1] + 1) \leq n$ many zeros. We can decode this bit sequence by reading the differences linearly because we know that $\text{LPnrF}[1] = 0$.

Algorithm 2: Determining the referred positions in a second pass described in Section 3.3.

1. $\lambda^R \leftarrow$ prev_leaf(prev_leaf(smallest_leaf)) and $\lambda \leftarrow$ smallest_leaf
2. clear B_V and set $B_V[\text{root node}] \leftarrow 1$
3. $W \leftarrow$ array of size $z \lg z$
4. **while** $\lambda \neq \lambda^R$ **do**
5.   **if** $B_L[\text{sufnum}(\lambda)] = 1$ **then**   ▷ turn of Player 1
6. $d \leftarrow 0$
7.   **while** $B_V[\text{level\_anc}(\lambda, d+1)] = 1$ **do** $d \leftarrow d+1$
8. $w \leftarrow \text{level\_anc}(\lambda, d)$
9.   **if** w is the root then output fresh factor
10.   **else** output referred position $W[B_W.\text{rank}_1(w)]$ ▷ invariant: $B_W[w] = 1$
11. $\lambda \leftarrow \text{suffixlink}(\lambda)$ ▷ end of Player 1's turn
12.   **foreach** node v on the path from λ^R up to the root **do**   ▷ turn of Player 2
13.   **if** $B_V[v] = 1$ **then break**   ▷ end turn on reaching an already marked node
14. $B_V[v] \leftarrow 1$
15.   **if** $B_W[v] = 1$ **then** $W[B_W.\text{rank}_1(v)] = 2n - \text{sufnum}(\lambda^R)$
16. $\lambda^R \leftarrow \text{prev\_leaf}(\lambda^R)$ ▷ end of Player 2's turn

Table 4. LPnrF and LPnF of our running example. Both arrays are defined in Section 4. See Section 5 for the definition of LPrF.

i	1	2	3	4	5	6	7	8	9	10	11
$T\#$	a	b	b	a	b	b	a	b	a	b	#
LPnrF	0	0	2	1	3	3	2	3	2	1	0
LPnF	0	0	1	3	3	3	2	3	2	1	0
LPrF	0	6	5	5	4	3	2	3	2	1	0

4.1. Adaptation of the Single-Pass Algorithm

Having an $\mathcal{O}(n)$-bits representation of LPnrF gives us hope to find an algorithm computing LPnrF in a total workspace space of $o(n \lg n)$ bits. Indeed, we can adapt our algorithm devised for the reversed LZ factorization to compute LPnrF. For that, we just have to promote all leaves to phrase leaves such that the condition in Line 7 of Algorithm 1 is always true. Consequently, Player 1 performs a root-to-leaf traversal for finding the lowest node marked in B_V of each leaf. By doing so, the time complexity becomes $\mathcal{O}(n^2)$, however, since we visit at most $\sum_{i=1}^{n} \text{LPnrF}[i] = \mathcal{O}(n^2)$ many nodes during the root-to-leaf traversals (there are strings like $T = \mathtt{a} \cdots \mathtt{a}$ for which this sum becomes $\Theta(n^2)$).

To lower this time bound, we follow the same strategy as in [22] ([Section 3.5]) or [34] ([Lemma 6]) using suffixlink and Lemma 2: After Player 1 has computed $\text{str\_depth}(w) = \text{LPnrF}[i-1]$ for w being the lowest ancestor marked in B_V of the leaf with suffix number $i-1$, we cache $\tilde{w} := \text{suffixlink}(w)$ for the next turn of Player 1 such that Player 1 can start the root-to-leaf traversal to the leaf $\tilde{\lambda}$ with suffix number i directly from $\tilde{w}$ and thus skips the nodes from the root to $\tilde{w}$. This works because $\tilde{w}$ is the ancestor of $\tilde{\lambda}$ with $\text{str\_depth}(\tilde{w}) = \text{LPnrF}[i-1] - 1$, and $\tilde{w}$ must have been marked in B_V since $\text{LPnrF}[i] \geq \text{str\_depth}(\tilde{w})$. See Figure 7 for a visualization. By skipping the nodes from the root to $\tilde{w}$, we visit only $\text{LPnrF}[i] - \text{LPnrF}[i-1] + 1$ many nodes during the i-th turn of Player 1. A telescoping sum together with Lemma 2 shows that Player 1 visits $\sum_{i=2}^{n}(\text{LPnrF}[i] - \text{LPnrF}[i-1] + 1) = \mathcal{O}(n)$ nodes in total.

The final bottleneck for CST are the n evaluations of $\text{str\_depth}(w)$ to compute the actual values of LPnrF (cf. Line 15 of Algorithm 1). Here, we use a support data structure on CST for str_depth [34] ([Lemma 6]), which can be constructed in $\mathcal{O}(\epsilon^{-1} n \log_\sigma^\epsilon n)$ time, uses $\mathcal{O}(n)$ bits of space, and answers str_depth in $\mathcal{O}(\epsilon^{-1} \log_\sigma^\epsilon n)$ time. This finally gives Theorem 2.

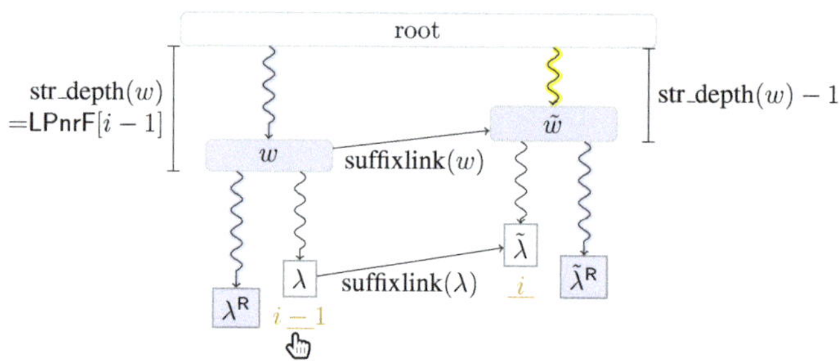

Figure 7. Setting of Section 4.1. Nodes marked in B_V are colored in blue (▢). Curly arcs symbolize paths that can visit multiple nodes (which are not visualized). When visiting the lowest ancestor of λ marked in B_V for computing $\mathsf{LPnrF}[i-1]$, Player 1 determines $\tilde{w} = \mathrm{suffixlink}(w)$ such that she can skip the nodes on the path from the root to the leaf $\tilde{\lambda}$ for computing $\mathsf{LPnrF}[i]$ (these nodes are symbolized by the curly arc highlighted in yellow (▢) on the right). There are leaves λ^R and $\tilde{\lambda}^R$ with suffix numbers of at least $2n - i + 2$ and $2n - i + 3$, respectively, since otherwise w would not have been marked in B_V by Player 2.

4.2. Algorithm of Crochemore et al.

We can also run the algorithm of Crochemore et al. [14] ([Algorithm 2]) with our suffix tree representations to obtain the same space and time bounds as stated in Theorem 2. For that, let us explain this algorithm in suffix tree terminology: For each leaf λ with suffix number i, the idea for computing $\mathsf{LPnrF}[i]$ is to scan the leaves for the leaf $\lambda^\star$ with $2n - \mathrm{sufnum}(\lambda^\star)$ being the referred position, and hence the string depth of $\mathrm{lca}(\lambda, \lambda^\star)$ is $\mathsf{LPnrF}[i]$. To compute $\lambda^\star$, we approach λ from the left and from the right to find λ_L (resp. λ_R) having the deepest LCA with λ among all leaves to the left (resp. right) side of λ whose suffix numbers are greater than $2n - i$. Then either λ_L or λ_R is $\lambda^\star$. Let $\ell_L[i] \leftarrow \mathrm{str\_depth}(\mathrm{lca}(\lambda_L, \lambda))$ and $\ell_R[i] \leftarrow \mathrm{str\_depth}(\mathrm{lca}(\lambda_R, \lambda))$. Then $\mathsf{LPnrF}[i] = \max(\ell_L[i], \ell_R[i])$, and the referred position is either $2n - \mathrm{sufnum}(\lambda_L)$ or $2n - \mathrm{sufnum}(\lambda_R)$, depending on whose respective LCA has the deeper string depth. Note that the referred positions in this algorithm are not necessarily always the leftmost possible ones.

Correctness. Let j be the referred position of the leaf λ with suffix number i such that $R[i\,..]$ and $R[2n - j\,..]$ have the LCP F of length $\mathsf{LPnrF}[i]$. Due to Lemma 1, there is a suffix tree node w whose string label is F. Consequently, λ and the leaf with suffix number $2n - j$ are in the subtree rooted at w. Now suppose that we have computed λ_L and λ_R according to the above described algorithm. On the one hand, let us first assume that $\ell_R[i] > \mathsf{LPnrF}[i]$ (the case $\ell_L[i] > \mathsf{LPnrF}[i]$ is treated symmetrically). By definition of $\ell_R[i]$, there is a descendant w' of w with the string depth $\ell_R[i]$, and w' has both λ_R and λ in its subtree. However, this means that $R[i\,..]$ and $R[\mathrm{sufnum}(\lambda_R)\,..]$ have a common prefix longer than $\mathsf{LPnrF}[i]$, a contradiction to $\mathsf{LPnrF}[i]$ storing the length of the longest such LCP. On the other hand, let us assume that $\max(\ell_L[i], \ell_R[i]) < \mathsf{LPnrF}[i]$. Then w is a descendant of the node w' being the LCA of λ and λ_R. Without loss of generality, let us stipulate that the leaf $\lambda^\star$ with suffix number $2n - j$ is to the right of λ (the other case to the left of λ works with λ_L by symmetry). Then $\lambda^\star$ is to the left of λ_R, i.e., $\lambda^\star$ is between λ and λ_R. Since $j > 2n - i$, this contradicts the selection of λ_R to be the closest leaf on the right hand side of λ with a suffix number larger than $2n - i$.

Finding the Starting Points. Finally, to find the starting points of λ_L and λ_R being initially the leaves with the maximal suffix number to the left and to the right of λ, respectively, we use a data structure for answering.

maxsuf_leaf(j_1, j_2) returning the leaf with the maximum suffix number among all leaves whose leaf-ranks are in $[j_1 \,..\, j_2]$.

We can modify the data structure computing max_sufnum in Section 3.3 to return the leaf-rank instead of the suffix number (the used data structure for max_sufnum first computes the leaf-rank and then the respective suffix number). Finally, we need to take the border case into account that λ is the leftmost leaf or the rightmost leaf in the suffix tree, in which case we only need to approach λ from the right side or from the left side, respectively.

The algorithm explained up to now already computes LPnrF correctly, but visits $\mathcal{O}(n)$ leaves per LPnrF entry, or $\mathcal{O}(n^2)$ leaves in total. To improve this bound to $\mathcal{O}(n)$ leaves, we apply two tricks. To ease the explanation of these tricks, let us focus on the right-hand side of λ; the left-hand side is treated symmetrically.

Overview for Algorithmic Improvements. Given we want to compute $\ell_R[i]$, we start with a pointer λ'_R to a leaf to the right of λ with suffix number larger than $2n - i$, and approach λ with λ'_R from the right until there is no leaf closer to λ on its right side with a suffix number larger than $2n - i$. Then λ'_R is λ_R, and we can compute $\ell_R[i]$ being the string depth of the LCA of λ_R and λ. If we scan linearly the suffix tree leaves to reach λ_R with the pointer λ'_R, this gives us $\mathcal{O}(n)$ leaves to process. Now the first trick lets us reduce the number of these leaves up to $2\ell_R[i]$ many for computing $\ell_R[i]$. The broad idea is that with the max_sufnum operation we can find a leaf closer to λ whose LCA is at least one string depth deeper than the LCA with the previously processed leaf. In total, the first trick helps us to compute LPnrF by processing at most $\sum_{i=1}^{n} \max(\ell_L[i], \ell_R[i]) = \mathcal{O}(n^2)$ many leaves. In the second trick, we show that we can reuse the already computed neighboring leaves λ_L and λ_R by following their suffix links such we process at most $2(\ell_R[i+1] - \ell_R[i] + 1)$ many leaves (instead of $2\ell_R[i+1]$) for computing $\ell_R[i+1]$. Finally, by a telescoping sum, we obtain a linear number of leaves to process.

First Trick. The first trick is to jump over leaves whose respective suffixes all share the same longest common prefix with $T[i\,..]$. We start with $\lambda_R \leftarrow$ maxsuf_leaf(leaf_rank$(\lambda) + 1$, $2n$) being the leaf on the right-hand side of λ with the largest suffix number. As long as sufnum$(\lambda_R) > 2n - i$, we search the leftmost leaf λ' between λ and λ_R (to be more precise: leaf_rank$(\lambda') \in$ [leaf_rank$(\lambda) + 1$ .. leaf_rank(λ_R)]) with lca$(\lambda', \lambda) =$ lca(λ_R, λ). Having λ', we consider:

- If leaf_rank$(\lambda') =$ leaf_rank$(\lambda) + 1$ (meaning λ' is to the right of λ and there is no leaf between λ and λ'), we terminate.
- Otherwise, we set λ'_R to the leaf with the largest suffix number among the leaves with leaf-ranks in the range [leaf_rank$(\lambda) + 1$ .. leaf_rank$(\lambda') - 1$]. If sufnum$(\lambda'_R) > 2n - i$, we set $\lambda_R \leftarrow \lambda'_R$ and recurse. Otherwise we terminate.

On termination, $\ell_R[i] =$ str_depth(lca(λ_R, λ)) because there is no leaf λ'' on the right of λ closer to λ than λ_R with str_depth(lca(λ'', λ)) $>$ str_depth(lca(λ_R, λ)) and sufnum$(\lambda'') > 2n - i$. Hence, sufnum(λ_R) is the referred position, and we continue with the computation of $\ell_R[i+1]$. See Figure 8 for a visualization.

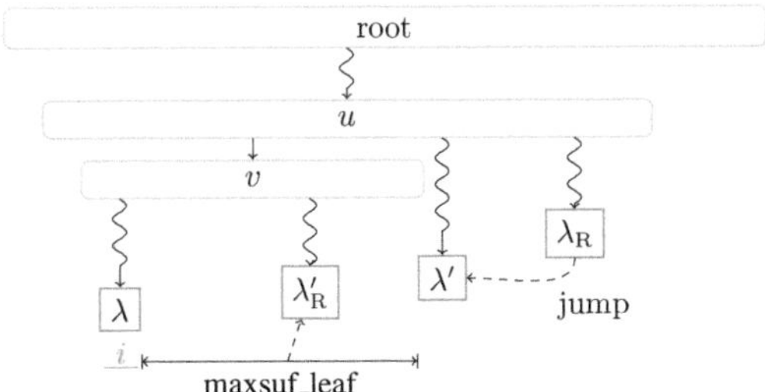

Figure 8. Computing LPnrF with [14] ([Algorithm 2]) as explained in Section 4.2. Starting at the leaf λ_R, we jump to the leftmost leaf λ' with lca$(\lambda', \lambda) =$ lca(λ_R, λ). Then, we use the operation max_sufnum$(\mathcal{I})$ returning the leaf-rank of the leaf λ'_R having the largest suffix number among the query interval $\mathcal{I} =$ [leaf_rank$(\lambda) + 1$ .. leaf_rank$(\lambda') - 1$]. If sufnum$(\lambda'_R) > 2n - i$, we recurse by setting $\lambda_R \leftarrow \lambda'_R$. The LCA of λ'_R and λ is at least as deep as the child v of u on the path towards λ (the figure shows the case that $v =$ lca(λ'_R, λ)), and hence $\ell_R[i]$ is at least str_depth(v) if we recurse.

Broadly speaking, the idea is that the closer λ_R gets to λ, the deeper the string depth of $\mathrm{lca}(\lambda_R, \lambda)$ becomes. However, we have to stop when there is no closer leaf with a suffix number larger than $2n - i$. So we first scan until reaching a λ' having the same lowest common ancestor with λ, and then search within the interval of leaves between λ and λ' for the remaining leaf λ'_R with the largest suffix number. We search for λ' because we can jump from λ_R to λ' with a range minimum query on the LCP array returning the index of the leftmost minimum in a given range. We can answer this query with an $\mathcal{O}(n)$-bits data structure in $\mathcal{O}(\epsilon^{-1})$ or $\mathcal{O}(\epsilon^{-1} \log_\sigma^\epsilon n)$ time for the SST or the CST, respectively, and build it in $\mathcal{O}(\epsilon^{-1}n)$ time or $\mathcal{O}(\epsilon^{-1}n \log_\sigma^\epsilon n)$ time (cf. [22] ([Section 3.3]) and [41] ([Lemma 3]) for details). However, with this algorithm, we may visit as many leaves as $\sum_{i=1}^n 2\ell_R[i] \le \sum_{i=1}^n 2\mathsf{LPnrF}[i]$ since each jump from λ_R to λ'_R via λ' brings us at least one value closer to $\ell_R[i]$. To lower this bound to $\mathcal{O}(n)$ leaf-visits, we again make use of Lemma 2 (cf. Section 4.1), but exchange $\mathsf{LPnrF}[i]$ with $\ell_R[i]$ (or respectively $\ell_L[i]$) in the statement of the lemma.

Second Trick. Assume that we have computed $\ell_R[i-1] = \mathrm{lca}(\lambda_R, \lambda)$ with $j :=$ sufnum$(\lambda_R) > 2n - i$. We subsequently set $\lambda \leftarrow \mathrm{suffixlink}(\lambda)$, but also $\lambda_R \leftarrow \mathrm{suffixlink}(\lambda_R)$. Now λ has suffix number i. If $\ell_R[i-1] \ge 1$, then the string depth of the $\mathrm{lca}(\lambda_R, \lambda)$ is $\ell_R[i-1] - 1$, and $R[\mathrm{sufnum}(\lambda_R) ..]$ is lexicographically larger than $R[\mathrm{sufnum}(\lambda) ..]$; hence λ_R is to the right of λ with sufnum$(\lambda_R) = j + 1$ (generally speaking, given two leaves λ_1 and λ_2 whose LCA is not the root, then $\mathrm{leaf\_rank}(\lambda_1) < \mathrm{leaf\_rank}(\lambda_2)$ if and only if $\mathrm{leaf\_rank}(\mathrm{suffixlink}(\lambda_1)) < \mathrm{leaf\_rank}(\mathrm{suffixlink}(\lambda_2)).$). Otherwise ($\ell_R[i-1] = 0$), we reset $\lambda_R \leftarrow \mathrm{maxsuf\_leaf}(\mathrm{leaf\_rank}(\lambda), 2n)$. By doing so, we assure that λ_R is always a leaf to the right of λ with sufnum$(\lambda_R) > 2n - i$ (if such a leaf exists), and that we have already skipped $\max(0, \ell_R[i-1] - 1)$ string depths for the search of λ_R with $\mathrm{str\_depth}(\mathrm{lca}(\lambda_R, \lambda)) = \ell_R[i]$. Since $\ell_R[i] \le \mathsf{LPnrF}[i]$, the telescoping sum $\ell_R[1] + \sum_{i=2}^n (\ell_R[i] - \ell_R[i-1] + 1) = \mathcal{O}(n)$ shows that we visit $\mathcal{O}(n)$ leaves in total.

In total, we obtain an algorithm that visits $\mathcal{O}(n)$ leaves, and spends $\mathcal{O}(\epsilon^{-1})$ or $\mathcal{O}(\epsilon^{-1} \log_\sigma^\epsilon n)$ time per leaf when using the SST or the CST, respectively. We need $\mathcal{O}(n)$ bits of working space on top of ST since we only need the values $\ell_L[i-1 .. i]$, $\ell_R[i-1 .. i]$, λ_L, and λ_R to compute $\mathsf{LPnrF}[i]$. We note that Crochemore et al. [14] do not need the suffix tree topology, since they only access the suffix array, its inverse, and the LCP array, which we translated to ST leaves and the string depths of their LCAs.

5. Open Problems

There are some problems left open, which we would like to address in what follows:

5.1. Overlapping Reversed LZ Factorization

Crochemore et al. [14] ([Section 5]) gave a variation of LPnrF that supports overlaps, and called the resulting array the longest previous reverse factor table LPrF, where $\mathsf{LPrF}[i]$ is the maximum length ℓ such that $T[i .. i + \ell - 1] = T[j .. j + \ell - 1]^R$ for a $j < i$. The respective factorization, called the overlapping reversed LZ factorization, was proposed by Sugimoto et al. [5] ([Definition 4]): A factorization $F_1 \cdots F_z = T$ is called the overlapping reversed LZ factorization of T if each factor F_x is either the leftmost occurrence of a character or the longest prefix of $F_x \cdots F_z$ that has at least one reversed occurrence in $F_1 \cdots F_x$ starting before F_x, for $x \in [1 .. z]$. We can compute the overlapping reversed LZ factorization with LPrF analogously to computing the (non-overlapping) reversed LZ factorization with LPnrF. As an example, the overlapping reversed LZ factorization of $T = \mathsf{abbabbabab}$ is $\mathsf{a} \cdot \mathsf{bbabba} \cdot \mathsf{bab}$. Table 4 gives an example for LPrF.

Since $\mathsf{LPrF}[i] \ge \mathsf{LPnrF}[i]$ by definition, the overlapping factorization seems more likely to have fewer factors. Unfortunately, this factorization cannot be expressed in a compact coding like Section 3.1 that stores enough information to restore the original text. To see this, take a palindrome P, and compute the overlapping reversed LZ factorization of $\mathsf{a}P\mathsf{a}$. The factorization creates the two factors a and $P\mathsf{a}$. The second factor is $P\mathsf{a}$ since $(P\mathsf{a})^R = \mathsf{a}P$. However, a coding of the second factor needs to store additional information about P to support restoring the characters of this factor. It seems that we need to store the entire left arm of P, including the middle character for odd palindromes.

Besides searching for an efficient coding for the overlapping reversed LZ factorization, we would like to improve the working space bounds needed for its computation. All algorithms we are aware of [5,14] embrace Manacher's algorithm [42,43] to find the maximal palindromes of each text position. To run in linear time, Manacher stores the arm lengths of these palindromes in a plain array of $n \lg n$ bits. Unfortunately, we are unaware of any time/space trade-offs regarding this array.

5.2. Computing LPF in Linear Time with Compressed Space

Having a $2n$-bit representation for four different kinds of longest previous factor tables (we can exchange LPnrF with LPrF in the proof of Lemma 2), we wonder whether it is possible to compute any of these variants in linear time with $o(n \lg n)$ bits of space. If we want to compute LPF or LPnrF within a working space of $\mathcal{O}(n \lg \sigma)$ bits, it seems hard to achieve linear running time. That is because we need access to the string depth of the suffix tree node w for each entry $\mathrm{LPF}[i]$ (resp. $\mathrm{LPnrF}[i]$), where w is the lowest node having the leaf λ with suffix number i and a leaf with a suffix number less than i (resp. greater than $2n - i$ for LPnrF) in its subtree, cf. [34] ([Lemma 6]) for LPF and the actions of Player 1 in Section 4.1 for LPnrF. While we need to compute str_depth(w) for determining the starting position of the subsequent factor (i.e., suffix number of the next phrase leaf, cf. Line 16) for the reversed LZ factorization, the algorithms for computing LPF (cf. [34] ([Lemma 6]) or [44] ([Section 3.4.4])) and LPnrF work independently of the computed factor lengths and therefore can store a batch of str_depth-queries. Our question would be whether there is a $\delta = \mathcal{O}((n \lg \sigma)/\lg n)$ such that we can accesses δ suffix array positions with a $\mathcal{O}(n \lg \sigma)$-bits suffix array representation in $\mathcal{O}(\delta)$ time. (We can afford storing δ integers of $\lg n$ bits in $\mathcal{O}(n \lg \sigma)$ bits.) Grossi and Vitter [45] ([Theorem 3]) have a partial answer for sequential accesses to suffix array regions with large LCP values. Belazzougui and Cunial [24] ([Theorem 1]) experienced the same problem for computing matching statistics, but could evade the evaluation of str_depth with backward search steps on the reversed Burrows–Wheeler transform. Unfortunately, we do not see how to apply their solution here since the referred positions of LPF and LPnrF have to belong to specific text ranges (which is not the case for matching statistics).

5.3. Applications in Compressors

Although it seems appealing to use the reversed LZ factorization for compression, we have to note that the bounds for the number of factors z are not promising:

Lemma 3. *The size of the reversed LZ factorization can be as small as $\lg n + 1$ and as large as n.*

Proof. The lower bound is obtained for $T = \mathtt{a} \cdots \mathtt{a}$ with $|T| = 2^{z-1}$ since $|F_1| = |F_2| = 1, |F_x| = 2|F_{x-1}|$ for $x \in [2 .. z]$ with $F_1 \cdots F_x = (F_1 \cdots F_x)^R$ being a (not necessarily proper) prefix of $T[|F_1 \cdots F_x| ..]$. For the upper bound, we consider the ternary string $T = \mathtt{abc} \cdot \mathtt{abc} \cdots \mathtt{abc}$ whose factorization consists only of factors of length one since $T^R = \mathtt{cba} \cdot \mathtt{cba} \cdots \mathtt{cba}$ has no substring of T of length 2 (namely, $\mathtt{ab}$, $\mathtt{bc}$, or $\mathtt{ca}$) as a substring (cf. [46] ([Theorem 5])). $\square$

Even for binary alphabets, there are strings for which $z = \Theta(n)$:

Lemma 4 ([46] (Theorem 9)). *There exists an infinite text T whose characters are drawn from the binary alphabet such that, for every substring S of T with $|S| \geq 5$, S^R is not a substring of T.*

Funding: This work is funded by the JSPS KAKENHI Grant Numbers JP18F18120 and JP21K17701.

Data Availability Statement: Not applicable.

Acknowledgments: We thank a CPM'2021 reviewer for pointing out that it suffices to store W in $z_W \lg n$ bits of space in Section 3.3, and that the currently best construction algorithm of the compressed suffix tree indeed needs $\mathcal{O}(\epsilon^{-1}n)$ time instead of just $\mathcal{O}(n)$ time.

Conflicts of Interest: The author declares no conflict of interest.

Appendix A. Flip Book

In this appendix, we provide a detailed execution of the algorithm sketched in Figures 4–6 by showing the state per turn and per player in Figures A1–A21. In our running example, each player has 10 turns.

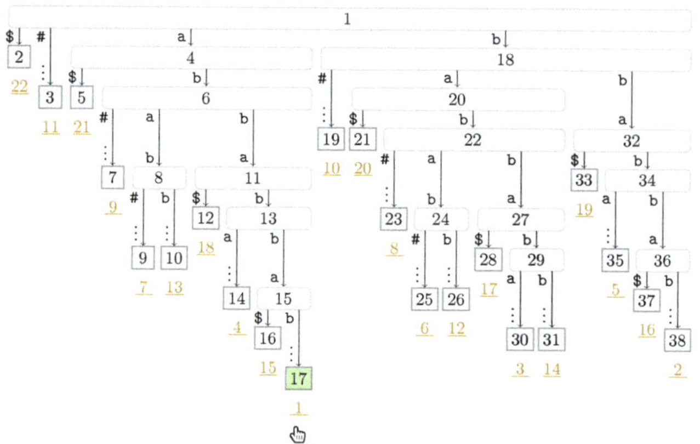

Figure A1. *Flip Book:* Initial State.

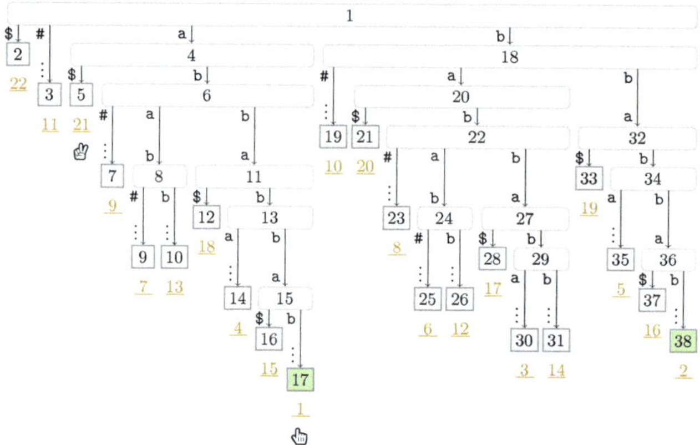

Figure A2. *Flip Book:* End of Turn 1 of Player 1.

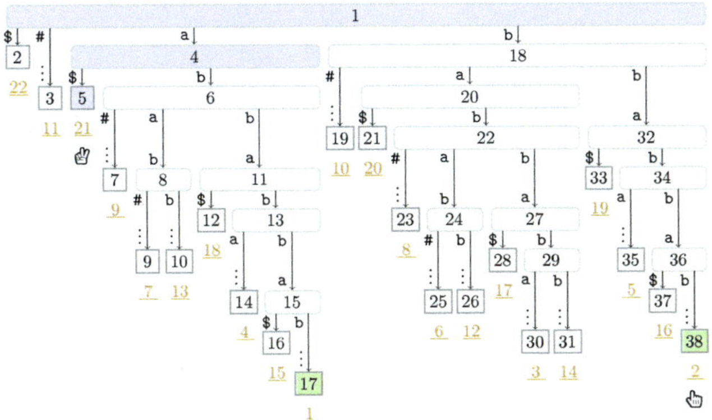

Figure A3. *Flip Book*: End of Turn 1 of Player 2.

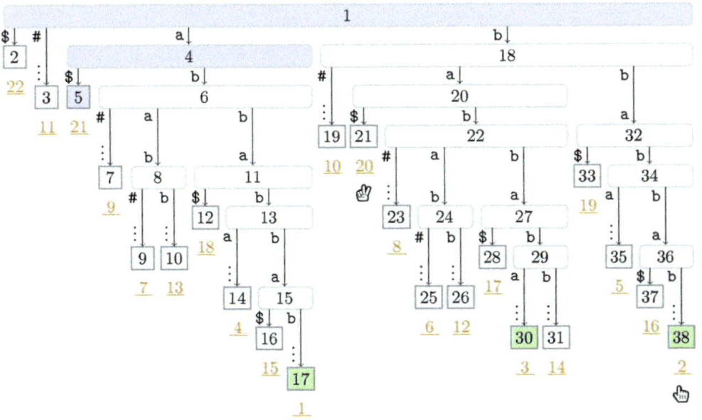

Figure A4. *Flip Book*: End of Turn 2 of Player 1.

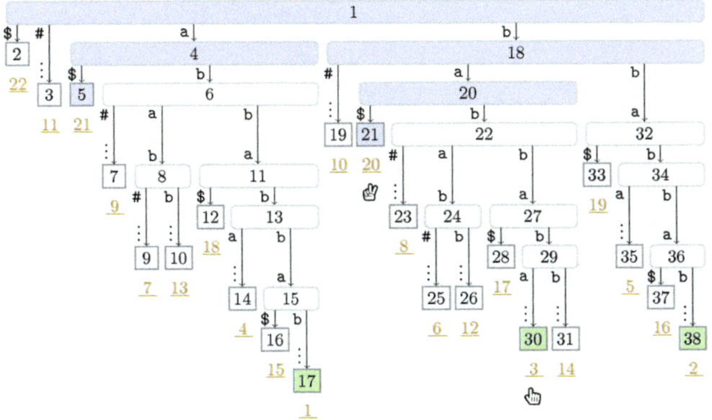

Figure A5. *Flip Book*: End of Turn 2 of Player 2.

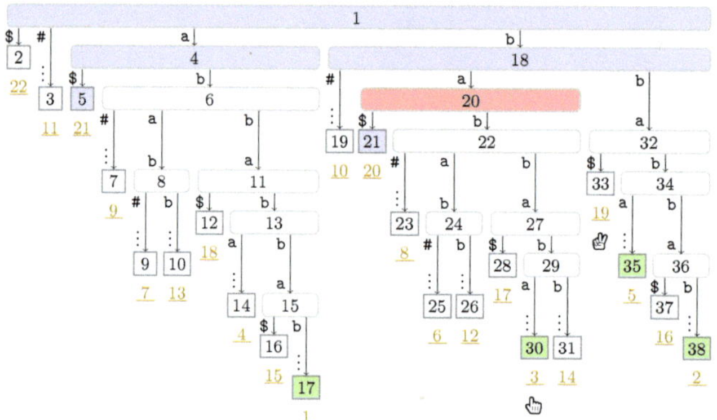

Figure A6. *Flip Book*: End of Turn 3 of Player 1.

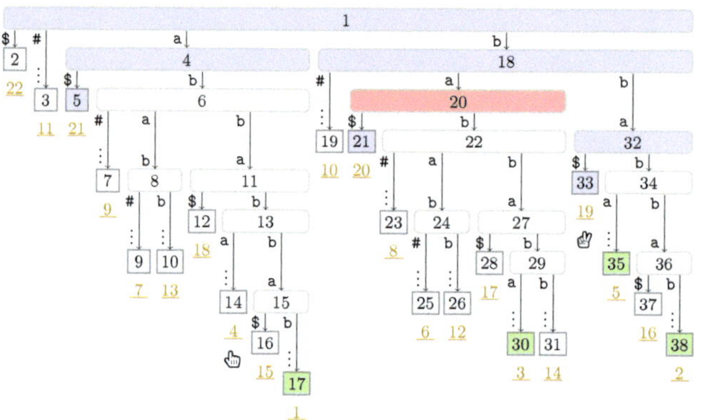

Figure A7. *Flip Book*: End of Turn 3 of Player 2.

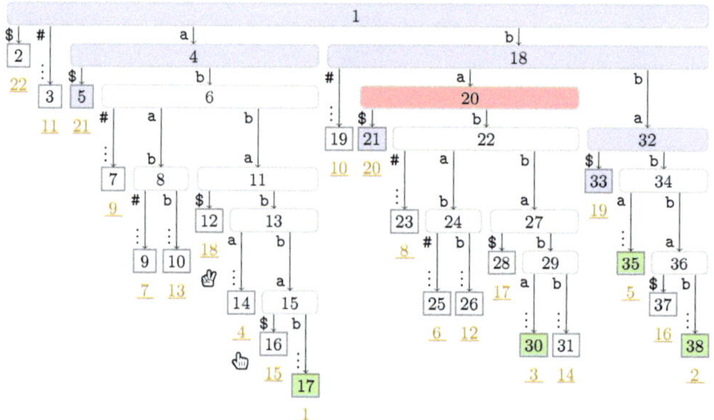

Figure A8. *Flip Book*: End of Turn 4 of Player 1.

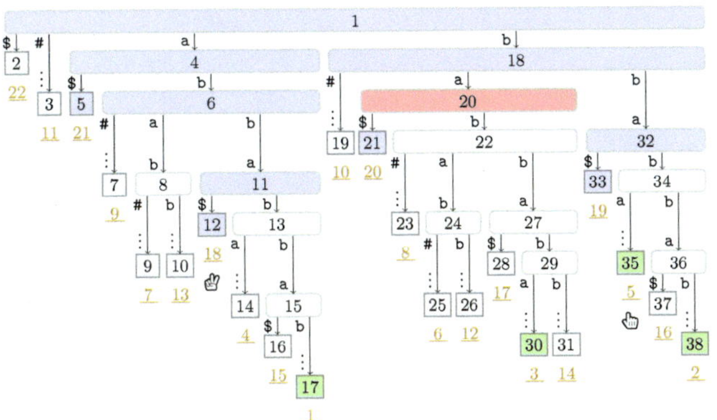

Figure A9. *Flip Book*: End of Turn 4 of Player 2.

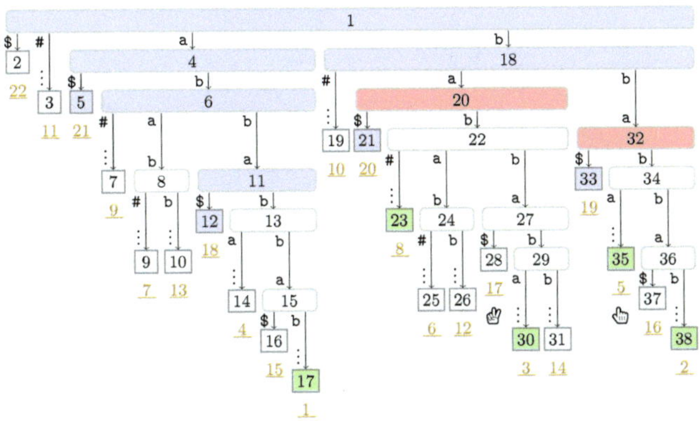

Figure A10. *Flip Book*: End of Turn 5 of Player 1.

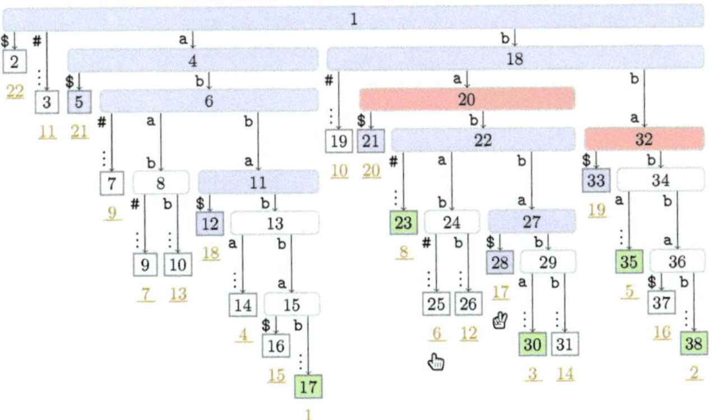

Figure A11. *Flip Book*: End of Turn 5 of Player 2.

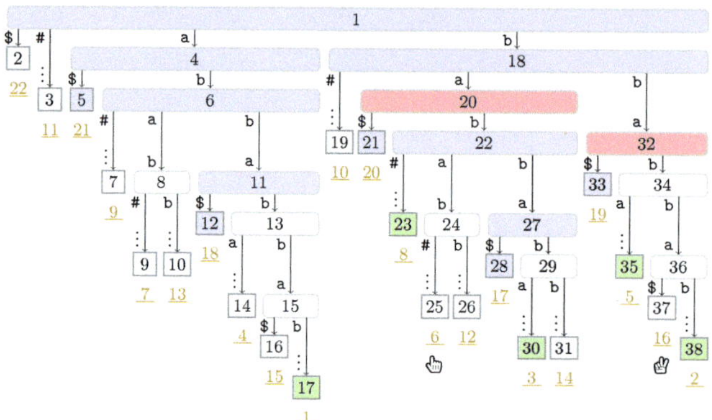

Figure A12. *Flip Book*: End of Turn 6 of Player 1.

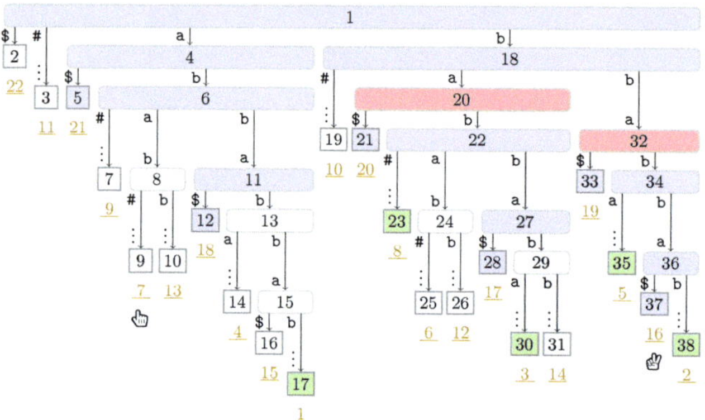

Figure A13. *Flip Book*: End of Turn 6 of Player 2.

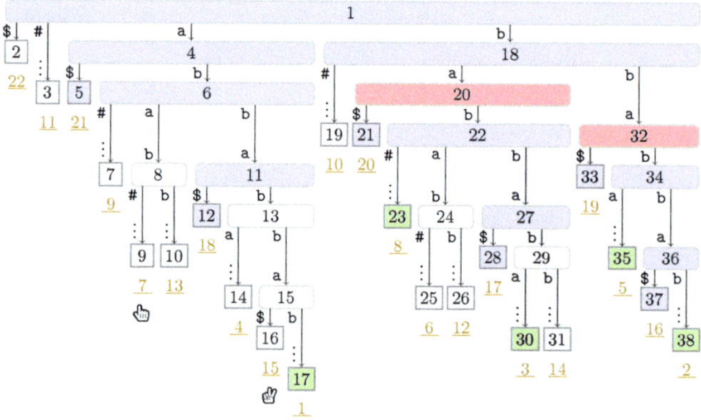

Figure A14. *Flip Book*: End of Turn 7 of Player 1.

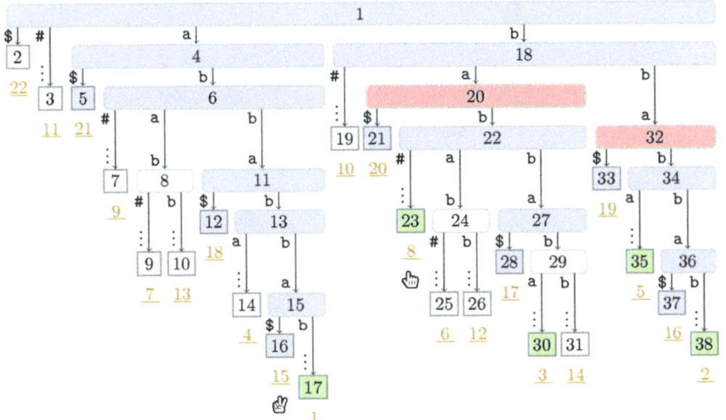

Figure A15. *Flip Book*: End of Turn 7 of Player 2.

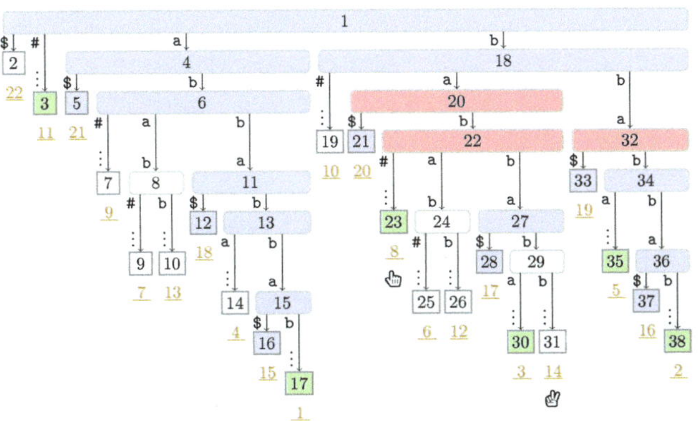

Figure A16. *Flip Book*: End of Turn 8 of Player 1.

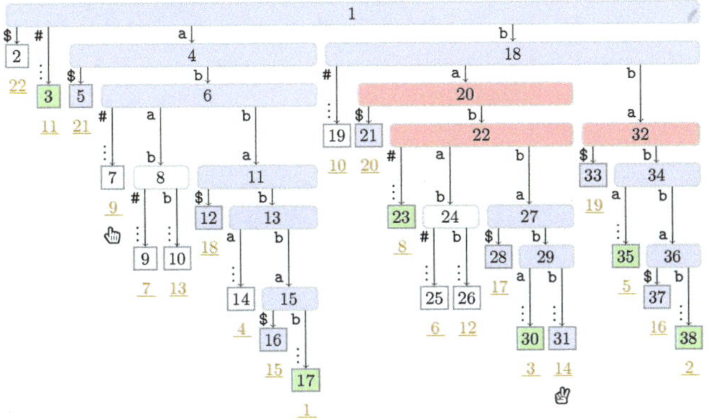

Figure A17. *Flip Book*: End of Turn 8 of Player 2.

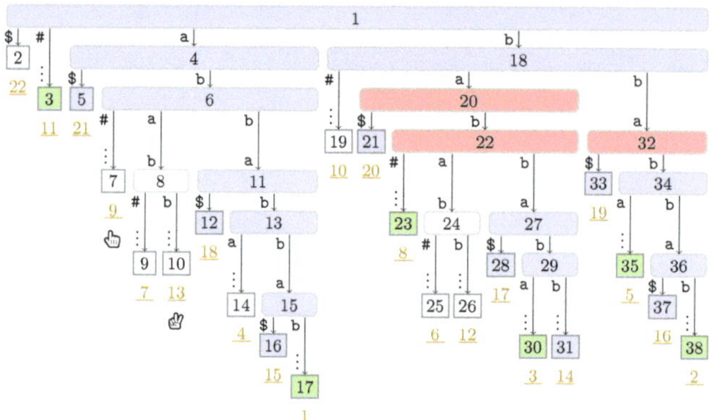

Figure A18. *Flip Book*: End of Turn 9 of Player 1.

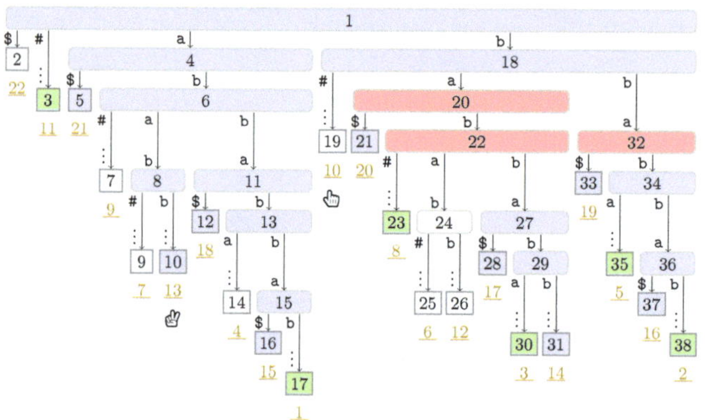

Figure A19. *Flip Book*: End of Turn 9 of Player 2.

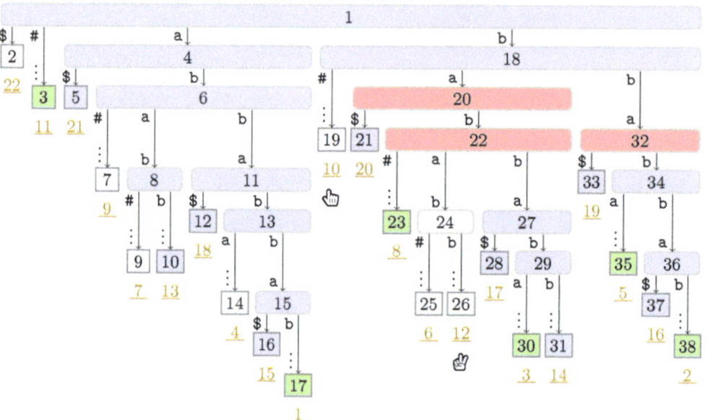

Figure A20. *Flip Book*: End of Turn 10 of Player 1.

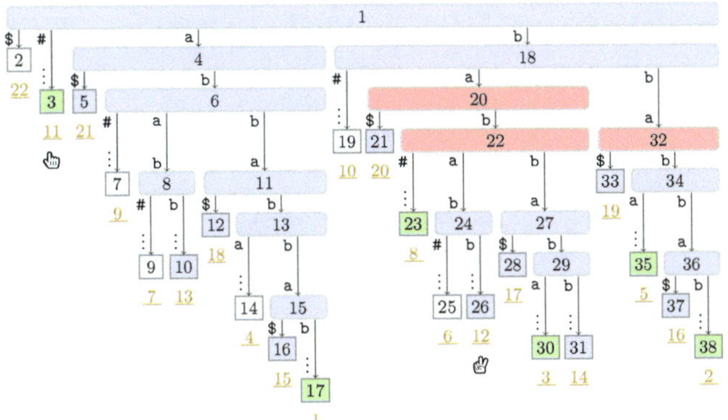

Figure A21. *Flip Book*: End of Turn 10 of Player 2.

References

1. Kolpakov, R.; Kucherov, G. Searching for gapped palindromes. *Theor. Comput. Sci.* **2009**, *410*, 5365–5373.
2. Storer, J.A.; Szymanski, T.G. Data compression via textural substitution. *J. ACM* **1982**, *29*, 928–951.
3. Crochemore, M.; Langiu, A.; Mignosi, F. Note on the greedy parsing optimality for dictionary-based text compression. *Theor. Comput. Sci.* **2014**, *525*, 55–59.
4. Weiner, P. Linear Pattern Matching Algorithms. In Proceedings of the 14th Annual Symposium on Switching and Automata Theory (swat 1973) SWAT, Iowa City, IA, USA, 15–17 October 1973; pp. 1–11.
5. Sugimoto, S.; Tomohiro, I.; Inenaga, S.; Bannai, H.; Takeda, M. Computing Reversed Lempel–Ziv Factorization Online. Avaliable online: http://stringology.org/papers/PSC2013.pdf#page=115 (accessed on 15 April 2021).
6. Chairungsee, S.; Crochemore, M. Efficient Computing of Longest Previous Reverse Factors. In Proceedings of the Computer Science and Information Technologies, Yerevan, Armenia, 28 September–2 October 2009; pp. 27–30.
7. Badkobeh, G.; Chairungsee, S.; Crochemore, M. Hunting Redundancies in Strings. In *International Conference on Developments in Language Theory*; LNCS; Springer, Berlin/Heidelberg, Germany, 2011; Volume 6795, pp. 1–14.
8. Chairungsee, S. Searching for Gapped Palindrome. Avaliable online: https://www.sciencedirect.com/science/article/pii/S0304397509006409 (accessed on 15 April 2021).
9. Charoenrak, S.; Chairungsee, S. Palindrome Detection Using On-Line Position. In Proceedings of the 2017 International Conference on Information Technology, Singapore, 27–29 December 2017; pp. 62–65.
10. Charoenrak, S.; Chairungsee, S. Algorithm for Palindrome Detection by Suffix Heap. In Proceedings of the 2019 7th International Conference on Information Technology: IoT and Smart City, Shanghai China, 20–23 December 2019; pp. 85–88.
11. Blumer, A.; Blumer, J.; Ehrenfeucht, A.; Haussler, D.; McConnell, R.M. Building the Minimal DFA for the Set of all Subwords of a Word On-line in Linear Time. In *International Colloquium on Automata, Languages, and Programming*; LNCS; Springer: Berlin/Heidelberg, Germany, 1984; Volume 172, pp. 109–118.
12. Ehrenfeucht, A.; McConnell, R.M.; Osheim, N.; Woo, S. Position heaps: A simple and dynamic text indexing data structure. *J. Discret. Algorithms* **2011**, *9*, 100–121.
13. Gagie, T.; Hon, W.; Ku, T. New Algorithms for Position Heaps. In *Annual Symposium on Combinatorial Pattern Matching*; LNCS; Springer, Berlin/Heidelberg, Germany, 2013; Volume 7922, pp. 95–106.
14. Crochemore, M.; Iliopoulos, C.S.; Kubica, M.; Rytter, W.; Walen, T. Efficient algorithms for three variants of the LPF table. *J. Discret. Algorithms* **2012**, *11*, 51–61.
15. Manber, U.; Myers, E.W. Suffix Arrays: A New Method for On-Line String Searches. *SIAM J. Comput.* **1993**, *22*, 935–948.
16. Dumitran, M.; Gawrychowski, P.; Manea, F. Longest Gapped Repeats and Palindromes. *Discret. Math. Theor. Comput. Sci.* **2017**, *19*, 205–217.
17. Gusfield, D. *Algorithms on Strings, Trees, and Sequences: Computer Science and Computational Biology*; Cambridge University Press: Cambridge, UK, 1997.
18. Nakashima, Y.; Tomohiro, I.; Inenaga, S.; Bannai, H.; Takeda, M. Constructing LZ78 tries and position heaps in linear time for large alphabets. *Inf. Process. Lett.* **2015**, *115*, 655–659.
19. Ziv, J.; Lempel, A. A universal algorithm for sequential data compression. *IEEE Trans. Inf. Theory* **1977**, *23*, 337–343.
20. Ziv, J.; Lempel, A. Compression of individual sequences via variable-rate coding. *IEEE Trans. Inf. Theory* **1978**, *24*, 530–536.
21. Fischer, J.; I, T.; Köppl, D.; Sadakane, K. Lempel–Ziv Factorization Powered by Space Efficient Suffix Trees. *Algorithmica* **2018**, *80*, 2048–2081.
22. Köppl, D. Non-Overlapping LZ77 Factorization and LZ78 Substring Compression Queries with Suffix Trees. *Algorithms* **2021**, *14*, 1–21.

23. Sadakane, K. Compressed Suffix Trees with Full Functionality. *Theory Comput. Syst.* **2007**, *41*, 589–607.
24. Belazzougui, D.; Cunial, F. Indexed Matching Statistics and Shortest Unique Substrings. In *International Symposium on String Processing and Information Retrieval*; LNCS; Springer: Cham, Switzerland, 2014; Volume 8799, pp. 179–190.
25. Franek, F.; Holub, J.; Smyth, W.F.; Xiao, X. Computing Quasi Suffix Arrays. *J. Autom. Lang. Comb.* **2003**, *8*, 593–606.
26. Crochemore, M.; Ilie, L. Computing Longest Previous Factor in linear time and applications. *Inf. Process. Lett.* **2008**, *106*, 75–80.
27. Belazzougui, D.; Cunial, F.; Kärkkäinen, J.; Mäkinen, V. Linear-time String Indexing and Analysis in Small Space. *ACM Trans. Algorithms* **2020**, *16*, 17:1–17:54.
28. Goto, K.; Bannai, H. Space Efficient Linear Time Lempel–Ziv Factorization for Small Alphabets. In Proceedings of the 2014 Data Compression Conference, Snowbird, UT, USA, 26–28 March 2014; pp. 163–172.
29. Kärkkäinen, J.; Kempa, D.; Puglisi, S.J. Lightweight Lempel–Ziv Parsing. In *International Symposium on Experimental Algorithms*; LNCS; Springer: Berlin/Heidelberg, Germany, 2013; Volume 7933, pp. 139–150.
30. Kosolobov, D. Faster Lightweight Lempel–Ziv Parsing. In *International Symposium on Mathematical Foundations of Computer Science*; LNCS; Springer: Berlin/Heidelberg, Germany, 2015; Volume 9235, pp. 432–444.
31. Belazzougui, D.; Puglisi, S.J. Range Predecessor and Lempel–Ziv Parsing. In Proceedings of the Twenty-Seventh Annual ACM-SIAM Symposium on Discrete Algorithms, Arlington, VA, USA, 10–12 January 2016; pp. 2053–2071.
32. Okanohara, D.; Sadakane, K. An Online Algorithm for Finding the Longest Previous Factors. In *European Symposium on Algorithms*; LNCS; Springer: Berlin/Heidelberg, Germany, 2008; Volume 5193, pp. 696–707.
33. Prezza, N.; Rosone, G. Faster Online Computation of the Succinct Longest Previous Factor Array. In *Conference on Computability in Europe*; LNCS; Springer: Cham, Switzerland, 2020; Volume 12098, pp. 339–352.
34. Bannai, H.; Inenaga, S.; Köppl, D. Computing All Distinct Squares in Linear Time for Integer Alphabets. Proc. CPM, 2017; Volume 78, LIPIcs, pp. 22:1–22:18. Available online: https://link.springer.com/chapter/10.1007/978-3-662-48057-1_16 (accessed on 16 April 2021).
35. Jacobson, G. Space-efficient Static Trees and Graphs. In Proceedings of the 30th Annual Symposium on Foundations of Computer Science Research, Triangle Park, NC, USA, 30 October–1 November 1989; pp. 549–554.
36. Clark, D.R. Compact Pat Trees. Ph.D. Thesis, University of Waterloo, Waterloo, ON, Canada, 1996.
37. Baumann, T.; Hagerup, T. Rank-Select Indices Without Tears. In Proceedings of the Algorithms and Data Structures—16th International Symposium, WADS 2019, Edmonton, AB, Canada, 5–7 August 2019; LNCS, Volume 11646, pp. 85–98.
38. Munro, J.I.; Navarro, G.; Nekrich, Y. Space-Efficient Construction of Compressed Indexes in Deterministic Linear Time. In Proceedings of the Twenty-Eighth Annual ACM-SIAM Symposium on Discrete Algorithms, Barcelona, Spain, 16–19 January 2017; pp. 408–424.
39. Burrows, M.; Wheeler, D.J. *A Block Sorting Lossless Data Compression Algorithm*; Technical Report 124; Digital Equipment Corporation: Palo Alto, CA, USA, 1994.
40. Lempel, A.; Ziv, J. On the Complexity of Finite Sequences. *IEEE Trans. Inf. Theory* **1976**, *22*, 75–81.
41. Fischer, J.; Mäkinen, V.; Navarro, G. Faster entropy-bounded compressed suffix trees. *Theor. Comput. Sci.* **2009**, *410*, 5354–5364.
42. Manacher, G.K. A New Linear-Time "On-Line" Algorithm for Finding the Smallest Initial Palindrome of a String. *J. ACM* **1975**, *22*, 346–351.
43. Apostolico, A.; Breslauer, D.; Galil, Z. Parallel Detection of all Palindromes in a String. *Theor. Comput. Sci.* **1995**, *141*, 163–173.
44. Köppl, D. Exploring Regular Structures in Strings. Ph.D. Thesis, TU Dortmund, Dortmund, Germany, 2018.
45. Grossi, R.; Vitter, J.S. Compressed Suffix Arrays and Suffix Trees with Applications to Text Indexing and String Matching. *SIAM J. Comput.* **2005**, *35*, 378–407.
46. Fleischer, L.; Shallit, J.O. Words Avoiding Reversed Factors, Revisited. *arXiv* **2019**, arXiv:1911.11704.

MDPI
St. Alban-Anlage 66
4052 Basel
Switzerland
Tel. +41 61 683 77 34
Fax +41 61 302 89 18
www.mdpi.com

Algorithms Editorial Office
E-mail: algorithms@mdpi.com
www.mdpi.com/journal/algorithms

9 783036 521947